Modeling the '50s

— The Glory Years of Rail —

Edited by *Model Railroader* staff

KALMBACH BOOKS

 Published by Kalmbach Publishing Co., 21027 Crossroads Circle, Waukesha, WI 53186.

This product is a Print on Demand format of the original book published by Kalmbach Publishing Company.

12 11 10 09 08   2 3 4 5 6

Cover photo: Paul J. Dolkos

With the exception of "America in the '50s," the material in this book has previously appeared in regular or special issues of *Model Railroader* magazine.

Publisher's Cataloging-In-Publication Data
(Prepared by The Donohue Group, Inc.)

Modeling the '50s : the glory years of rail / edited by Model Railroader staff.

p. : ill. ; cm.

With the exception of "America in the '50s," the material has previously appeared in regular or special issues of Model Railroader magazine.
ISBN: 978-0-89024-718-1

1. Railroads--Models--History--20th century. 2. Railroads--Models--Design and construction. I. Title: Modeling the fifties II. Title: Model railroader.

TF197 .M63 2008
625.1/9

CONTENTS

By Dick Christianson

It's about the memories

What was so great about the 1950s? What does such a large percentage of the model railroading community find so fascinating about the decade after WWII? I've thought a lot about these questions, so let me offer a few reminiscences that might provide some answers.

Grade crossings

It was always on a Friday night. My dad and his two brothers would get together to watch the fights – in black and white, of course. Wives, sons, and daughters played Clue or Sorry in the kitchen.

Finally, it would be time to go home, and we'd pile into our '53 two-tone cream-and-copper Chevy. About halfway home, we'd often get stopped at the grade crossing on the Madison Street hill by a Chicago & North Western freight out of Minneapolis bound for Chicago. Mom and I would count the cars; I especially liked the boxcars with the goat on the side. I still try to remember the motive power: a big Northern? Or maybe A-B-B-A F units in the C&NW's freight scheme? Could have been yellow-and-green striped Geeps, but I don't think so.

The clothesline

Or it was a summer weekday. As soon as she heard the Ten-Wheeler's whistle, Mom would throw open the screen door and hurry out to pull the freshly washed clothes off the line. Wet or dry – it didn't matter. If she didn't get them down, the sheets and underwear would be freckled with black spots of soot.

Smokey, our appropriately named black cocker spaniel, would chase the black behemoth down the cinder-ballasted right-of-way until it was gone or until the engineer, with a grin on his face, released a shot of steam at the brave (or foolish) little black dog. I've often wondered what Smokey thought he was doing – silly dog.

Streamliners (and airplanes)

Saturday afternoons were good too. Dad would open the Biscayne's door and ask if I wanted to go to the depot and watch the *400* pull in. *Was he kidding?* We'd park the car and walk across the unevenly brick-paved platform just as the spectacular yellow, silver, and green streamliners rounded the curve. Dad would take my hand, so I wouldn't cross the platform's yellow warning stripe.

The majestically sloped nose of the lead E unit carried a Mars light that had that hypnotic figure-8 motion. We'd watch the train get closer and then roll to a silent stop in front of the station. After a few minutes, the conductor called "All aboard," closed the lower half of the door, and looked out as the train effortlessly and without a sound began to roll toward Altoona, Merrillan, Milwaukee, and Chicago. (Later on, Dad would drive me to the airport to watch North Central Airlines' DC-3s land – a sign of the changing times.)

A ride in the cab

Summer vacations were best. We could ride our bikes to the depot or up the Third Street hill to stand on the bridge and watch freights or the *400* cross the high bridge over the Chippewa River. Or, if we were really adventurous, my good buddy Dick Martinson and I would ride our bikes to Altoona to visit his grandpa, Arne Anderson, and maybe get a tour of the roundhouse there.

Arne worked for the Omaha Road as foreman of the bridge crew out of Altoona. Once we managed to get a short cab ride across the turntable and down the roundhouse lead and back. That was nearly 50 years ago, and I remember it as though it were yesterday.

Still swell

So, why are the 1950s so fascinating? In my case, it's the memories. And for some reason, a lot of my memories involve trains – big, powerful, gritty, dramatic, interesting, fast, and noisy. Oh, sure, our HO or N versions of the 1950s are small, mostly quiet, and, well, they're models. But they help us remember times when we were wide-eyed and innocent. We didn't know rail passenger service was declining and that being a railroader was dangerous work. All we knew was that trains were, to use the vernacular … swell! And, in my case – and maybe yours – because of the memories, they still are.

Finding the '50s

By Tony Koester

Half a century has passed, but if you know where to look and who to talk to, reminders of the 1950s are still out there. Those of us who lived through the '50s have very vivid memories. Who could forget Elvis, rock 'n' roll, pink-and-gray suede shoes, hardtops, or steam's final curtain?

By and large, the 1950s were a magical time. Perhaps that's why we strive to depict the highlights of those years on our model railroads.

Putting the '50s in context

You can't model something convincingly without understanding what makes it tick. To appreciate the brightly colored first-generation diesel paint schemes, you need to imagine a nation shrugging off the rationing of World War II.

The blacks, tans, and grays of pre- and postwar autos gave way to bold, almost brash, two-tone schemes and chrome – lots of chrome. My dad bought a 1956 Ford hardtop painted orange over white. If that's not enough visual excitement for you, how about hot pink over white?

By the end of the 1950s, the postwar renaissance was wearing thin. The railroads had abandoned the gaiety espoused in the original covered-wagon paint schemes for somber, and especially cheaper, "dip" schemes. For example, the Baltimore & Ohio went from yellow-trimmed blue and gray diesels to solid-blue units with spartan B&O lettering. The magic was fading.

Rekindling the magic

Thanks to model railroading, we can turn the clock back and resurrect the magic of the '50s. As-delivered diesel paint schemes were still in vogue, and passenger trains still played a role in our transportation needs.

On some roads, steam was on the way out; others, notably the coal-hauling Norfolk & Western, believed steam had miles to go before it slept. The Nickel Plate ordered the last 10 of its fabled fleet of Berkshires in 1949, keeping the diesel salesmen at bay until the recession of 1958.

Discovering exactly when diesels bumped steam aside usually isn't as simple as reading a date in a book. It helps to understand not just when but how the transition took place on the railroad you're modeling.

Those lucky enough to come across dispatchers' train sheets from their railroad and period of interest can form an accurate portrait. Most railroad historical societies (listed on www.ModelRailroader.com) have such materials in their archives, and members often have private collections. Railroadiana shows are also a good source.

I spend many enjoyable hours on field trips with friends gathering information. We photograph and measure structures and interview railroaders and industry managers. We visit libraries, antique shops, historical societies, and county surveyor offices.

At conventions, you're likely to meet modelers who share your interests and professional railroaders who can tell you how it was. However, workers from the '50s are now well past retirement age, so if you don't ask questions now, their knowledge may be lost.

The Nickel Plate Road ran both diesels and steam locomotives east out of Frankfort, Ind., until 1956. But on the St. Louis Line, steam vanished almost overnight in July 1955 when GP9s like no. 452 arrived. Tony Koester photo

Employee time books

Employee time books are an overlooked source of key operational data. Each crewman kept a record of the trains he worked: date, train number, locomotive, starting and ending points, mileage (which influenced pay), often the tonnage and conductor's name, and occasionally the caboose number. Time books were such a part of a railroader's life that they were often saved, making them among the most accessible and accurate documents you can obtain. You can find time books at antique shops, on online auction sites, and through the assistance of a professional railroader, who could check with former colleagues.

The media age

Advances in technology have increased access to research materials. The Internet can be a portal to a gold mine of railroad resources. Old amateur movies have been transferred to videotapes or DVDs. And priceless collections of Kodachromes from the 1940s and '50s are being reproduced in books, so we can now see the color of the equipment and structures and how they weathered.

We can discern the colors of signs, vehicles, and people's clothing. We note that highway centerline stripes were usually white, not yellow, and are reminded that mailboxes were olive-green and stop signs were yellow.

Living in the '50s

At an antique show, I bought a coral-colored radio that reeked of the 1950s. Through its speaker, I can pipe the sounds of the '50s into the crew lounge. Magazines for off-duty crews to browse through are all from fall 1954 or earlier. When crews operate on my depiction of the NKP, I want them to feel they've ridden a time machine back to the '50s. Take it from one who was there: It's worth the trip!

1-1

CHAPTER ONE

America in the '50s

By Robert S. McGonigal

In an era before strip malls and "big box" stores, Americans did their shopping at storefront commercial districts like this stretch of Grand Avenue in Chicago, where soon-to-be-replaced streetcars mingled with streamlined buses and automobiles. CTA photo, Krambles-Peterson Archive

The United States today is very different from the nation it was at the peak of the railroad age a century ago. In many ways, the 1950s were the "hinge decade" between two eras, a time when many great changes began or accelerated.

Before World War II, when automobile ownership was far from universal, urban development tended to be dense and centralized. People lived near their places of employment and shopping districts – or along railroad or transit lines that could take them there, **1-1**. In and around cities, residential streets were laid out in grid patterns with small lots, multistory department stores were the place to shop, and the skyscraper was symbolic of corporate success. In small towns, Main Street or the courthouse square, not a highway strip at the edge of town, was the center of commercial and social activity.

Especially in the East, many cities and towns existed before railroads came to them in the 19th century. In such cases, railroads tried to establish stations in the downtown area. Where geographic or operating considerations made this impractical, the railroads got as close as they could.

They themselves became magnets for development, prompting the establishment of satellite commercial districts around their stations. When railroads built across "unsettled" territory, they often founded towns every few miles as sources of future traffic. Befitting their origin, these communities tended to be laid out with the depot as the central element.

Although these land-development patterns had their roots in the 19th century, they were still very much in evidence in the 1950s. However, they were under assault by key elements of the prosperity that followed World War II: sharp increases in automobile ownership and highway mileage. Though both had been rising for decades, in the postwar era, they combined with a desire of many Americans to forsake city life for country living – or at least a version of it that came to be known as suburbia.

Throughout the 1950s, farmland and forests around cities were subdivided into residential lots at an unprecedented pace. Curvilinear street patterns and ranch houses gained favor. Shopping centers, then office and industrial parks – many with no rail service – followed the subdivisions to the suburbs.

In small cities and towns, the ascendancy of the auto was seen in the development of commercial strips along highways at the edge of town. These too had been germinating since the 1920s, but they really took off in the '50s, by which time they included service stations (predecessors of today's pump-it-yourself gas-and-gos), motels, drive-in theaters, car dealerships, diners, and other auto-oriented land uses.

Another victim of automobiles (and buses) was the streetcar. Although they had been in decline since World War I, trolleys still plied the streets of dozens of American cities in 1950. A decade later, fewer than 10 systems remained, all but two of which (Johnstown, Pa., and New Orleans) used streamlined PCC-type cars exclusively.

The not-so-open roads

When blissfully tooling down an old highway today, it's easy to forget that in the 1950s the experience might not have been so pleasant. Road development lagged behind auto use, with the result that many two- and four-lane city streets and rural highways were frequently clogged with traffic. The situation was especially bad in cities and towns, where streets were overwhelmed by commuters and cross-country travelers who, lacking the option of beltways or bypasses, had no choice but to fight their way through the heart of town.

On the "open road," it was common to see a semi-truck slowly grinding up a hill with a dozen or more cars forced to crawl along behind until the truck crested the grade. Then it was a race to see how many cars could get by before the next double yellow line appeared.

Thousands of miles of new urban expressways and intercity superhighways cut through the landscape in the '50s. The rapid-transit right-of-way in the median of the new Congress Expressway in this Chicago view was unusual – most of the new roads ignored rail. T. H. Desnoyers photo, Krambles-Peterson Archive

1-2

1-3

As automobile ownership soared in the 1950s, more and more land was needed for parking lots, such as this one at the New York City Transit System's Willets Point Road station. George Krambles photo, Krambles-Peterson Archive

America had no grade-separated, long-distance express highways until 1940, when the Pennsylvania Turnpike opened between points near Harrisburg and Pittsburgh. The "dream highway" was an instant success, but World War II put other such projects on hold.

After the war, turnpikes in northeastern states made it possible to travel from New York to Chicago without encountering a single traffic light – a godsend to motorists accustomed to battling the likes of U.S. 30 between the nation's two biggest cities. In a few metropolitan areas like New York and Los Angeles, urban expressways relieved city streets of their traffic burden.

Superhighway expansion shifted into high gear in 1956 when President Eisenhower signed a bill authorizing the construction of a 41,000-mile network of new roads – the Interstate Highway System. Termed "the greatest engineered structure in the world," the Interstate system soon began revolutionizing travel by bringing standardized on-ramps, cloverleafs, overpasses, and mile after mile of intersection-free four-lane divided highway to all corners of the nation, **1-2**.

Despite the start of work on the Interstate system, most highways of the 1950s were two-lane, at-grade affairs. Most had been built during the preceding 40 or 50 years, and many directly paralleled railroad lines. The two-lane concrete highway – punctuated with yellow (yes, yellow) stop signs, lattice-based billboards, and post-and-wire guardrails – beside a railroad track is as much an icon of the 1950s as the big Buicks, DeSotos, and Nashes that *thump-a-thumped* along it.

Indeed, the '50s were perhaps the golden age of the automobile. Rarely, before or since, have cars been so attention-getting or symbolic of status, freedom, and personal identity. With enormous chromed grilles and bumpers, jet-like tail fins, jaunty two-tone finishes, and all-around presence, cars of the '50s were forces to be reckoned with, **1-3**. During the '50s, the driving vacation became an institution – passenger trains hardly stood a chance against the flexibility and economy of the family car.

Popular culture and conflict

Surpassing the car as a 1950s cultural force was television. At the start of the decade, only about 10 percent of the nation's households owned a television set; at the end, 86 percent of American homes had an antenna on the roof or rabbit ears in the living room. Suburban neighborhoods looked like they were topped by a forest of metallic trees!

Though some predicted that television would bring about the end of motion pictures, Hollywood created some of its most memorable films in the '50s, including *High Noon* (1952), *Rear Window* (1954), and *Ben Hur* (1959), and created lots of young stars like James Dean, Marlon Brando, and Marilyn Monroe. While most theaters were still downtown (malls with mega theaters weren't part of the scene yet), drive-in theaters (a.k.a. "passion pits") were popular with teens and less-so with their parents (in separate vehicles, of course).

Lurking behind the postwar movement to suburbia and its accompanying optimism was the dark possibility that it all could end in an instant. When the Soviet Union exploded its first atomic bomb in 1949, it was clear that America and its post-World War II rival were engaged in a game of unimaginably high stakes. As nuclear-capable missiles, bombers, and submarines proliferated, civilians fought the Cold War with backyard bomb shelters and "duck-and-cover" air raid drills. Yellow air raid shelter signs were evident everywhere, and rooftop spotters watched for Soviet bombers flying over the Arctic to attack our cities.

There were other aspects to the "rockin'-good" '50s. With more than a third of the workforce employed in manufacturing, far more Americans were engaged in making things than today, when just 11 percent work in manufacturing. Large industrial structures and heavy equipment to gather raw materials were prominent on the landscape – and invariably connected by rail lines.

Food was just beginning to get "fast" with TV dinners and restaurants promising to make it easy for the American family to dine in or out. Hundreds of orange-roofed Howard Johnson's restaurants set the standard for consistent, high-quality food from coast to coast. And when the first McDonald's opened in 1955, few could have foreseen the golden arches' eventual supremacy.

1-4

Five decades ago, T-shirts and flip-flops were strictly for the beach, and people dressed up when they traveled. Linn H. Westcott photo

Among all of the titanic changes in America during the 1950s was one that went largely unnoticed: the slipping of railroads from the nation's consciousness. Of course railroading was not entirely forgotten, but in a forward-looking age of fast food, flashy cars, jet airplanes, and new ways of living, trains increasingly came to represent the past.

As people turned to automobiles and airplanes, railroad passenger service dried up, causing railroads to slip further from the public eye, **1-4**. Steam locomotives, especially, marked railroading as an antiquated institution. Yet, when they disappeared, so did the most visible and compelling aspect of the industry.

Even though by the end of the decade, IBM and Boeing had replaced the Pennsylvania and Southern Pacific as companies with cachet, railroading remained a fascinating, important part of the American scene in the 1950s, especially to youngsters – today's baby boomers – who were there to witness the change. How else can one explain the fact that the 1950s remain – in the first decade of the 21st century – model railroading's most popular decade?

CHAPTER TWO

Modeling the SP in Los Angeles

By Robert Smaus

2-1

Filled with unique buildings, downtown Los Angeles anchors one end of my 1950s-era Southern Pacific layout. The building in the back gets its outrageous green color from glazed terra-cotta tiles. It's loosely modeled on the 1929 Eastern Building. Photos by the author

Sometimes things like commercial buildings or a geographic region inspire the choices a modeler will make when building a layout; other times it's a railroad's rolling stock and locomotives. For me it was the latter – those bright orange Pacific Fruit Express (PFE) refrigerator cars.

I model the Southern Pacific in Los Angeles of the 1950s. In the downtown area, I wanted to model some of the interesting brick buildings built around the turn of the century, **2-1**. Tracks were everywhere, slicing through buildings and often ending up inside some.

Because I wanted to run long blocks of orange reefers behind big Cab-Forwards, besides the LA cityscape, I also needed to model orange groves, packing plants, icing platforms, and even a box manufacturer to supply the picking boxes and orange crates (see chapter 13). All the industries on my layout actually existed in Southern California in the '50s, though not always where I located them.

I filled my layout with freight cars that I saw in photos, so sometimes I had to make very specific cars – scratchbuilding, using resin kits, or kitbashing them to match those of the prototype.

My model yard combines several SP yards found in or near downtown LA, including the Cornfield and mighty Taylor. Taylor Yard was where the PFE icing docks were located, so reefers from my town of Arroyo are iced here before heading to points east.

Although my plan is a looped dog-bone style layout, it can be run as a point-to-point railroad. Trains run from the yard to Arroyo and return, or they run from the yard to downtown LA and back. When enough cars are collected in the yard, I make up a good-size train and send it east. Usually this means it makes several trips around the layout before ending up on storage tracks behind the backdrop. While the trains go round and round, I sit back and enjoy the freight cars rolling by. There's no point building 'em if you can't watch 'em!

With that said, let's take a tour of the layout. We're bound to see some interesting freight cars and industries along the way, which might make good candidates for your own modeling projects.

Layout at a glance

Name: Southern Pacific Los Angeles Division
Scale: HO (1:87.1)
Size: 12 x 19 feet
Prototype: Southern Pacific
Locale: Southern California
Era: early 1950s
Style: walkaround
Mainline run: 85 feet
Minimum radii: 33" main line, 28" sidings, 24" industrial spurs
Minimum turnouts: no. 6 main line, no. 5 sidings
Maximum grade: none
Benchwork: L girder
Height: 54"
Roadbed: HomaBed and Homasote sheets
Track: code 83 flextrack
Scenery: plaster over aluminum screen
Backdrop: painted hardboard
Control: DC cab control

2-2

The diesel fueling stand shown here was typical for the 1950s. The small orange cart was used to carry supplies to the cabooses on the next track.

2-3

I made my mill from a combination of Walthers, Design Preservation Models, and Stewart Products kits. Those Kadee Buffalo Creek flour cars are ready for pickup.

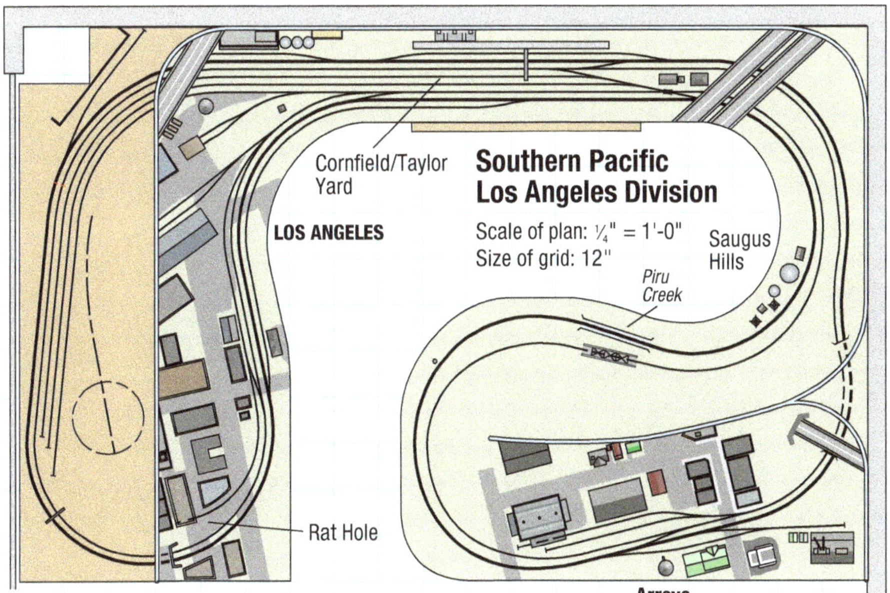

Illustrations by Jay Smith

2-4

To model chalk marks that had been rubbed off, I used dry transfers and applied white and gray powdered pastels to the car sides.

2-5

By the 1950s, the sign on the inbound freight shed had fallen off, but it left a paint shadow, which I dutifully modeled on my building.

2-6

***Overnight* cars were seldom, if ever, seen in any other trains, so without the sheds, there would be no reason to have them on the layout.**

The Cornfield Yard

Cars generated by Capitol Milling and other industries concentrated at Cornfield Yard. This double-ended yard fed cars across the Los Angeles River to Taylor Yard, which was the main yard for Los Angeles.

ENGINE TERMINAL

A crew's day starts at a small engine terminal tucked into the corner of the Cornfield Yard. This terminal is used for switch engines and caboose storage. By the mid-1950s, most of the yard switchers were Electro-Motive Division SW1 diesels, although some steam switch engines were still in use.

I built a diesel fueling stand by piecing together bits of plastic pipe, brass rod, and pieces of insulation stripped from a wire, **2-2**. The rough asphalt paving is painted Sculptamold. I gussied up the Walthers SW1 switcher with a Cannon & Co. cab.

THE CORNFIELD

Grain spilled on the way to Capitol Milling gave the Cornfield its name (formally it was part of River Station Yard). Sections of the old mill were built at different times and of different materials, **2-3**.

On the cars, chalk marks tell yard crews where the car is headed, **2-4**. Mine were made from dry transfers by Clover House.

River Station freight house

Inside a curve, I found room for the River Station freight sheds, where those famous *Overnight* cars would be found. This was also the site where the first SP piggyback trains departed.

FREIGHT SHEDS

On the way to the warehouse district, you pass the old River Station freight sheds. These were located inside a wye off the Cornfield, so curving track on both sides is believable. There were two sheds, inbound and outbound, and each was more than a block long; I modeled only the end of the inbound one, **2-5**.

In the '50s, the sheds were used for the SP's *Overnight* trains that ran between LA and San Francisco. I keep several Sunshine models of the black *Overnight* cars on the truncated sidings, **2-6**.

2-7

Wearing bright *Daylight* colors, an SP trailer waits on a loading ramp. The White 3000 cab-over tractor is an Ivers Engineering kit.

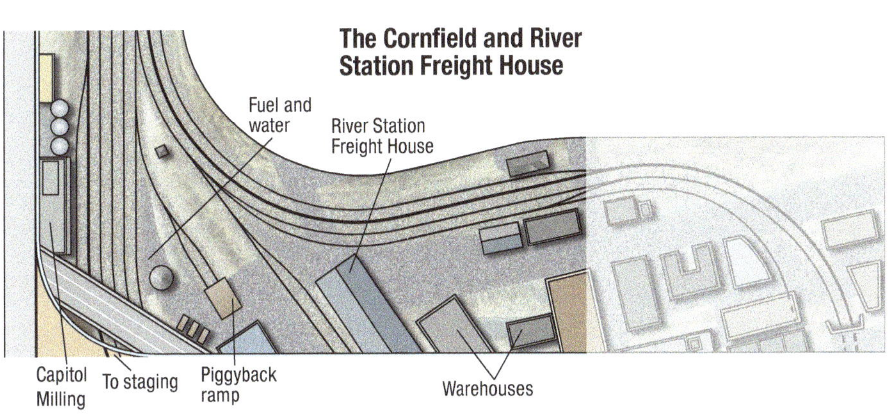

PIGGYBACK SERVICE

Where the first SP piggyback trains departed, cars were loaded circus-style using a simple concrete ramp. I carefully measured the prototype ramp and then built the plaster model.

The SP-owned trucks were painted in glorious *Daylight* colors, **2-7**. I made my colorful trailers by wrapping sheet styrene around Athearn trailer bodies.

Alameda Street

Alameda Street runs diagonally for seven feet before fading into the backdrop. It makes the street seem to go on forever and allows space for many buildings, giving the modeled area a downtown flavor. By placing the tallest buildings at one end of the street, near the front of the layout, I created a false perspective that makes the street appear even longer. The commercial alley behind Alameda, where the Star Warehouse is located, extends four feet.

2-8

Sidings curved off Alameda Street to the businesses along it, and the trains often snarled traffic.

TRACKS ALONG ALAMEDA

The Southern Pacific's double-track main line ran right down the middle of busy Alameda Street, with sidings curving off to the many businesses that lined it, **2-8**. I had to put my tracks alongside Alameda for lack of space, but they still run by period businesses such as the Walnut Exchange and Aggeler & Musser Seeds. Two treasured Magnuson resin structures that I've moved from layout to layout now reside on Alameda Street.

2-9

A Santa Fe extended-height boxcar and an extra-low Boston & Maine car sit on a siding in front of a Magnuson resin structure adorned with a huge O scale Lucky Strike dry transfer.

WAREHOUSE BOXCARS

Some of my favorite cars can be found on these sidings, **2-9**. One is a Santa Fe class Bx12 extended-height car (a resin

2-10

With its yellow, green, and white livery, a GMC municipal bus, along with a variety of period businesses, helps set the scene and time. Partially visible on the Angelus Hotel is a large RKO sign that I made from dry transfer letters.

2-11

Built of flexible sheet styrene, the Davies Warehouse features a curved wall with two trackside doors.

2-12

Even though it's a rectangular building, two sides of the Star Warehouse can't be seen, so they have no detail.

2-13

This Baldwin VO1000 (a Stewart model) features trolley poles that were used to trip crossing signals.

kit made by Westerfield), which probably carries furniture. Another favorite car is an extra-low Boston & Maine car. This resin kit, made by Yankee Clipper, is now available from Funaro & Camerlengo.

BUSINESSES ON ALAMEDA

This street is lined with familiar businesses of the time including Grant's Department Store, Owl Drugs, Dr. Beauchamp, the Hotel Cecil, and the Angelus Hotel, **2-10**. I made the giant painted sign (copied from an old photo) advertising RKO Radio Pictures on the Angelus by using dry transfer letters.

The Rat Hole

One particularly tight curve leads to a convoluted siding off Alameda near Second Street that was called the Rat Hole. I modeled the beginning of this siding where the tracks curved around Davies Warehouse and then went under and though Star Warehouse. On my layout, the siding ends there, but in LA, it kept winding between buildings for blocks. Star was built over the tracks and Davies inside the curve's radius because real estate downtown was valuable. No space was wasted.

I partly kitbashed and partly scratchbuilt these models. Their construction was featured in the May 2004 and June 2006 issues of *Model Railroader*. Unfortunately, these brick buildings were damaged during the 1987 Whittier earthquake and subsequently knocked down.

DAVIES WAREHOUSE

Davies Warehouse was a distinctive commercial building with a curved wall of loading doors parallel to the tracks on one side. It was nestled into an odd-shaped lot inside a curved spur. My model is modified to fit a similarly odd-shaped space on my layout, **2-11**. Because of the curve, engines ran right up against the sides of the buildings. Signs that said "No Clearance for Man on Side of Car" weren't kidding. I located my warehouse close enough to the track to make that true!

STAR WAREHOUSE

The Star Warehouse, with a footprint of 6¼" × 15¼", is much larger than Davies. However, on my layout, it's merely a background structure. Only the covered trackside loading dock is visible from the aisle, **2-12**. It has an elaborate Romanesque portico that leads to a covered loading platform.

TROLLEY POLES

It wasn't uncommon to see a Baldwin VO1000 switcher lettered for the Pacific Electric subsidiary running on SP tracks, **2-13**. Some PE locomotives were equipped with trolley poles, not for power, but to trip crossing signals.

On the prototype, the poles hang over the handrails at the front and rear by about a foot. (To build this locomotive, see *Model Railroader* October 2001.)

Taylor Yard

Taylor Yard was the main yard for Los Angeles. Everything funneled across a long, narrow bridge, which was originally double-tracked but later carried a single track.

I allowed room for icing tracks and a seven-car icing platform next to the yard. A good portion of the actual Taylor Yard was full of icing platforms and belonged to the PFE.

The yard tracks run straight for 10 feet before going under the Broadway Street Bridge and curving out of sight.

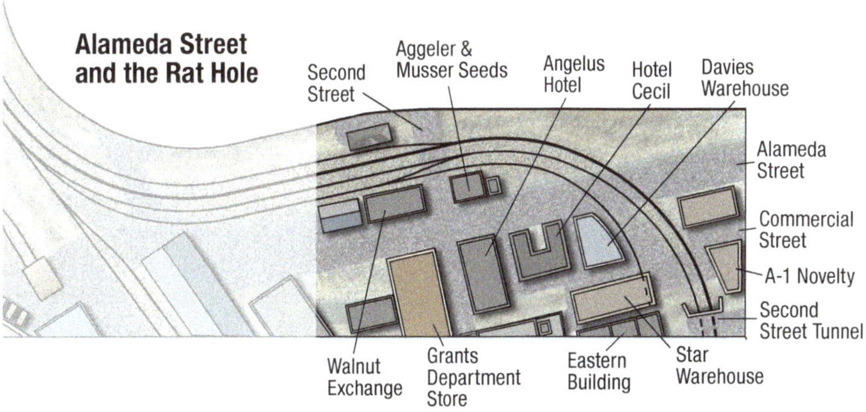

Even though these PFE cars look superficially alike, they vary in detail quite a bit.

I modeled the Dayton Avenue Tower using layers of sheet styrene. A switcher waits in the shade of the overpass.

2-17

Behind the train is a Central Valley rigid truss bridge kit that was based on the actual Piru Creek Bridge. However, I had to cut a section out of it, so it would fit in the short straight section of an S curve on my main line.

2-18

I modeled the oil fields in the Saugus Hills using a few old N.J. International drilling rigs and some pumps from S.S. Ltd. and Diamond Scale.

2-19

I based this Sunkist packing plant on one in Fillmore and used art board and balsa to build it.

2-20

General Box Distributors provides the crates used by Sunkist for packing its oranges.

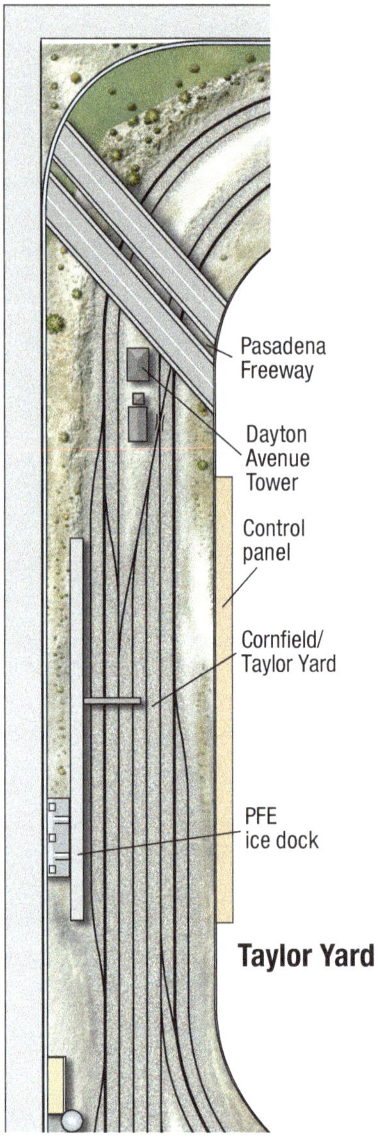

THE YARD

Freight cars picked up downtown are brought back to the yard and switched into trains headed east. Half of my yard models the Cornfield, but the other half actually models the much bigger Taylor Yard.

To get the feel of this big yard, I combined the two yards into one and made it pass though the backdrop, under the Broadway Street Bridge, **2-14**.

ICING PLATFORM

I made the icing platforms by combining several Creative Model Associates kits. The finished model is five feet long and accommodates a respectable nine refrigerator cars.

I had fun modeling a variety of cars from PFE's huge fleet,. Through the years, there were wood-sheathed and all-steel cars, variations in paint schemes, and the addition of fans and other upgrades, **2-15**. Fortunately, I've found a number of accurate models for these cars in both resin and plastic.

DAYTON AVENUE TOWER

As trains follow the tracks under the Pasadena Freeway, they pass Dayton Avenue Tower, an imposing cast concrete structure, **2-16**. I modeled it using layers of sheet styrene. Switchers often waited in the shade of the overpass.

Piru Creek and the Saugus Hills

This location is the best place to watch long trains as they snake though the S curve in front of the hills. There is little here except desert scenery and one lone SP tool house, so this scene gives a feeling of the vast empty distances traveled between LA and Arroyo.

PIRU CREEK

On the way to Arroyo, the SP line crosses Piru Creek. Since the Central Valley rigid truss bridge kit was actu-

ge, the atex nd in

vere At red id a very

t

rus ed st e

YOUR RECEIPT
THANK YOU
23:40 07-25-2019
MC NO.0000 4896
$19.95T1
SUBTOTAL $19.95
TAX1 $1.66
TOTAL-TAX $1.66
TOTAL $21.61
CASH $21.61
HAVE A NICE DAY
PLEASE COME AGAIN

SUNKIST PLANT

I pretty much designed my model railroad around a citrus packing plant that I'd made well before I'd started the layout, **2-19**. It's big, and the only place it would fit was inside a turnback curve. Unfortunately, the real structure lost its handsome Mission-style covered loading area and the front facade in a 1987 earthquake.

2-21 Houses such as this are common to the region. I modeled all the houses by kitbashing Rix Products kits.

BOX FACTORY

Right next door to the Sunkist plant is a box manufacturer that supplies packing crates, **2-20**. Since lumber was shipped in boxcars, inside gondolas, or on flatcars, you never know what type of cars you'll find sitting on this siding.

MORE OF ARROYO

A small residential neighborhood sits behind the commercial buildings, **2-21**. At the far end of town, there's a large feed mill that gets boxcars and early covered hoppers periodically. And if the switching crew has time, they can grab a bite to eat at a small diner called the Spot Cafe before returning to LA.

3-1

CHAPTER THREE

Setting the scene

By Paul J. Dolkos

Dwight Eisenhower is running for president, an advertisement for a 1953 Chevy is being put up, the trees are bare, and the weeds are dead. These are just some of the details I used on my HO scale Boston & Maine layout to convey to visitors and operators that it's November 1952.

Photos by the author

Prototype accuracy is something many of us strive for on our layouts. However, we can quickly miss that goal if we have items on our model railroad that do not fit the era we're re-creating. For example, a Franklin Roosevelt campaign billboard would be as out of place on a layout set in 1952 as would an Electro-Motive Division GP35 on a 1959-themed layout.

As modelers, we need to remember that the era we're modeling may not be evident to all visitors. Therefore, we need to use details to help people recognize that our model railroad is set during a specific time in a specific place, **3-1**. I'll show how I used details to subtly convey that it's November 1952 on my Boston & Maine layout set in northern New Hampshire.

Recalling the '50s through photos

It is difficult remembering what the world looked like a few years ago, let alone a half-century ago. To get an idea of what things looked like back then, I like to study photos from that period.

Books are probably the best resource to find images. One I found especially helpful was LeRoy O. King's book *Streetcar Scenes of the 1950s in Color* published by Morning Sun Books.

It is also important to broaden your research beyond railroading titles. I've found that books containing 1950s news photographs and publications on period architecture have given me a better understanding of the look and feel for the era that I'm modeling.

The World Wide Web is also an excellent resource for finding period photographs. Many historical organizations have photo archives that are searchable online, and some groups even sell copies of the images in their collections.

When you look at the photographs, identify items that make strong visual references to the 1950s. Some things you'll want to pay particular attention to are vehicles, businesses, architectural styles, and clothing. The objective is to incorporate as many clues as possible to the period on your layout. By carefully choosing these elements, you'll clearly convey to operators and visitors what year you're trying to re-create.

Mid-century styles

Figures are usually small enough that precise details aren't obvious, but we can paint our figures in clothing styles common for the 1950s. In the gas station scene on my layout, **3-2**, I have an adult man wearing a trench coat, white dress shirt, black pants, and a tie. The Gulf gas station attendants are wearing light brown uniforms and caps.

Women of the '50s wore long dresses and coats, while railroad employees wore blue or gray bib overalls and traditional cloth caps. Studying photographs is a good way to see how people dressed. The style was conservative, and the colors were much more basic than they are today.

Signs of the time

Signs are an effective way to let visitors know both what year you're modeling and the locale of your layout. On the road into my town of Woodsriver, N.H., there are two billboards, **3-3**. One is a campaign advertisement for Dwight Eisenhower, suggesting the year is either 1952 or 1956. The other sign is promoting a 1953 Chevrolet, narrowing the time frame to late 1952.

I also use road signs appropriate for the era I'm modeling. For example, the stop signs have yellow backgrounds with black letters. Yellow signs were typical until 1954, when national highway standards were changed and the red-and-white stop sign was adopted.

Retail advertising signs also help convey that it's 1952 on my layout. Sign shapes, fonts, and lighting methods evolved throughout the decade, particularly on retail establishments.

There were also more regional gas stations, department stores, and beverages during the 1950s. Incorporating these names on the storefronts and billboards on your layout will help your

3-2

Clothing is one clue that helps identify a time. In this gas station, two uniformed attendants service a man's car. Another clue is the sign advertising Gulf premium No-Nox for 25 cents a gallon.

3-3

It's November 1952 as evidenced by two billboards situated in fall foliage. One carries an election message, and in the other, a '53 Chevy is replacing a '52 Ford.

layout's theme come across more clearly to visitors and operators.

Along the line

Railroad equipment and lineside details are two more ways to help viewers know what year and what area you're modeling. In 1952, the Boston & Maine rostered an interesting blend of steam and diesel locomotives, **3-4**. The B&M didn't run steam in northern New Hampshire in 1952, but the handful of steam locomotives on my layout did operate elsewhere on the railroad. If a visitor asks why I'm operating steam, I tell them dieselization was delayed by a few years.

3-4

In 1952, the Boston & Maine's roster contained both diesels and steam locomotives. Here a BL2 rolls past Woodsriver on well-groomed roadbed.

3-5

The remains of a water tower foundation can be seen in the staging yard. This detail suggests that the steam-to-diesel transition is taking place.

Lineside details can also offer clues as to the year. In my staging yard, the foundation is all that remains of a water tower that supplied water to steam locomotives, **3-5**. Other details appropriate for the era include yard switch stands with lamps and semaphore signals. In the mid-20th century, the roadbed also had a different appearance, **3-4**. Ballast had sharp edges and was laid on a bed of cinders.

Architecture change

Even though structures often predate the era we're modeling, you can still detail them in ways that suggest it's the 1950s. Over its lifetime, a structure may change radically. Trim and cupolas are removed, modern additions are built, and new paint is added.

During the 1950s, as well as in later decades, many older structures had metal facades applied. Brick buildings were still being constructed, but they were plain with little or no ornamentation, **3-6**.

The appearance of factories changed during the mid-20th century as well. New processes and equipment forced companies to expand or renovate their operations. For example, a 1940s-era paper mill might have featured all-brick structures. A decade later, metal-sheathing may have replaced or joined the masonry.

How these industries were served also changed. In the 1940s and early '50s, boxcars were used to transport pulpwood. As new unloading terminals were constructed, pulpwood cars became commonplace at paper mills.

3-6

Buildings along this brick-paved street include several plain brick structures and one built of wood, which contains an interesting group of advertising signs (at left).

Layout room decor

You can further reinforce the era and locale of your model railroad with room decorations. For example, I have this 1952 calendar hanging on the wall in my layout room.

Other items you could display include prototype photos, posters, newspapers, and commercial signs.

I've even thought about having a sign painted that reads "Welcome to northern New Hampshire, November 1952" in the stairwell leading to the layout room. Such an item would set the stage for operators and help visitors better understand what I'm trying to do.

CHAPTER FOUR

Adding vintage trucks – and realism

By Lou Sassi

4-1

A variety of trucks populated the highways during the 1950s. Adding these trucks to your layout gives it a feeling of authenticity.

Photos by the author

Whether it is a semi-trailer at a warehouse loading dock, a pickup at a construction site, or a hi-rail vehicle parked next to the depot, trucks add realism to a layout, **4-1**. Trucks and trains just seem to go together.

The number of 1950s-era truck models, offered as kits or ready-to-run, has increased greatly over the past decade. Most of the truck kits are either cast resin or white metal, and both kinds require some skill to assemble. Ready-to-use vehicles are, as the name suggests, assembled, painted, and can go from the package to your layout with little or no extra work.

No matter if the truck is a kit or ready-to-run, by adding a few details you can really make your vehicles stand out.

All the models described in this chapter are HO scale. While truck models are available in several scales, the selection in 1:87 proportion is much greater, which makes it easier to show the different classes of trucks from the 1950s.

Semi-trailers

If a company wasn't served by rail or if it had goods that didn't require shipment by railroad, it would use semi-trailers. On your layout, semi-trailers would look appropriate at a warehouse, transloading center, or truck terminal.

Mack's B61 tractor was a common sight on highways throughout the United States in the 1950s. Athearn offers a ready-to-use B61 tractor, **4-2**, and Alloy Forms has one in kit form. The prototype B61, and the firm's B20 through B80 series tractors, were first produced in 1953 and stayed in production for the next 14 years.

For behind the B61 tractor, Classic Metal Works provides a line of injection-molded plastic trailers in both N and HO scales. The trailers are available decorated for a variety of trucking companies as well as undecorated. One example is a Freuhauf 32-foot trailer decorated for Eastern Motor Express, **4-2**, which includes both raised and lowered landing gears that a modeler can install.

One manufacturer, Ulrich, has been producing truck models for more than a half century. The company's original line of tractors and trailers, like their prototype counterparts, was first produced in the 1950s. Most of the firm's products have been reissued. A full list of Ulrich products can be found on its Web site, www.ulrichmodels.com.

When modeling local prototypes, adding details provides a finishing touch. As an example, George Micklus made a semi-trailer based on a prototype owned by Nuzzi Oil Co. in northern New Jersey, **4-3**. George started with an Ulrich Mack Bulldog H60 metal tractor kit and added several details, including windshield wipers, horn, headlight lenses, and marker light jewels. He made the window glazing from clear styrene sheet cut to fit.

International Harvester, a well-known farm implement manufacturer, also produced tractors during the 1950s. Classic Metal Works offers single-axle, ready-to-use, die-cast metal R-190 tractors in N and HO scales, **4-4**. The firm's models are based on the prototype produced from 1953 to 1968.

Dual-wheeled trucks

To help better distribute weight and increase the payload of trucks, manufacturers placed four wheels on a single

4-2

This Athearn ready-to-roll HO scale Mack B61 tractor is seen here with an Eastern Motor Express Freuhauf 32-foot trailer from Classic Metal Works.

4-3

George Micklus built this Mack Bulldog H60 cab-over-engine tractor from an Ulrich metal kit. He added an assortment of detail parts to the semi-trailer to match the prototype used by a New Jersey oil dealer.

4-4

An exhaust stack and other details have been added to this IH R-190 tractor from Classic Metal Works. Tractors are available in decorated and undecorated versions.

4-5

This 1955 Chevrolet two-ton stakebed truck is produced by Alloy Forms. The vehicle, sold as a metal kit, features a single rear axle with dual wheels.

4-6

Mack offered its B-series trucks with an assortment of body styles, including a bulk-fuel tank, as shown on this model from an Alloy Forms kit.

4-7

Fire trucks feature dual wheels on the rear axle in order to carry more weight. Alloy Forms manufactures a Mack B42/B61 pumper truck kit with an open cab, as shown here, or with a closed cab.

4-8

White Motor Co. produced the 3000-series of cab-over-engine vehicles from 1949 until the late 1950s. The John B. Stetson Fine Hats truck shown here is from an out-of-production Ivers Engineering kit.

axle instead of two. The dual-wheel design was used on an assortment of industrial trucks.

One such vehicle was the stakebed truck. Using a kit from Alloy Forms, I built a 1955 Chevrolet two-ton truck with a stakebed body and decorated it for Green Acres Farm, **4-5**, but these trucks were by no means just farm vehicles. Stakebed trucks were used by tire dealers, railroads, and other businesses to transport large items.

Another truck with a dual-wheel design is a fuel-delivery tanker truck, **4-6**. Many homes and businesses used heating oil in the fall and winter months, making a delivery truck an ideal vehicle for a layout set in the northern half of the United States.

Fire trucks, such as the Mack B42/B61 pumper, **4-7**, would look great parked in front of a fire house. This model is also by Alloy Forms and is available with an open or closed cab.

White Motor Co. of Cleveland, Ohio, produced its 3000-series of cab-over-engine vehicles from 1949 to 1957, **4-8**. Classic Metal Works offers White cab-over-engine tractors with an assortment of body styles, including box-style delivery truck, dump truck, fuel truck, and tractor. The ready-to-use die-cast metal vehicles feature detailed cab interiors.

Pickup trucks

The pickup truck was, and still is, one of the most common vehicles on the road. Railroads, construction firms, and many other businesses use pickups to transport tools and other supplies to and from job sites.

Photo **4-9** features three HO pickup models that will fit well on a 1950s layout. They include a Dyna Models' 1947 Ford with window glazing, license plates, and a Preiser driver; a 1950 Chevy, manufactured by Busch, that I converted into a phone utility truck by adding storage racks and a ladder; and a 1956 Ford from Alloy Forms, which Frank Pisante assembled, painted, and detailed.

Milk delivery trucks

Door-to-door milk delivery was common in the 1950s, so a milk truck parked in front of a house or driving through a neighborhood would look appropriate. Jack Wright built the Magnuson (now Wm. K. Walthers) model that roams the streets of Adams on my West Hoosic Division layout, **4-10**. Woodland Scenics also offers an HO scale milk truck in its AutoScenes line.

George Bacon, chief engineer at the Detroit Electric Vehicle Co., designed the electric milk truck in 1922. When Detroit Electric passed on the idea, Bacon enlisted some investors to start the Detroit Industrial Vehicle Co. (DIVCO). The DIVCO milk trucks were powered by a gasoline engine.

In 1937, the DIVCO milk truck was redesigned with an all-steel welded van body and a snub nose. This body style was used until January 1986, and a few can still be found in service today.

Track maintenance

If too many of the trucks on your layout are competing with rail traffic, why not make a few of them into railroad service trucks?

Rich Cobb spliced an Alloy Forms B61 Mack tractor cab onto a Roco tank retriever body to make a Boston & Maine HO scale service truck, **4-11**. He put two HO scale wheelsets and a 55-gallon barrel on the flatbed, added N scale wheels to simulate hi-rail wheels, and painted the vehicle in B&M colors. The vehicle would look appropriate next to the main line as part of a maintenance-of-way scene or parked near a depot or in a yard.

4-9

Pickup truck models from left to right are a 1947 Ford (Dyna Models kit), a 1950 Chevy (modified Busch ready-to-use vehicle), and a 1956 Ford (Alloy Forms kit).

4-10

Wm. K. Walthers produces a line of resin vehicle kits, including a milk truck based on a Borden's Detroit Industrial Vehicle prototype. Jack Wright built this model, which delivers milk on my West Hoosic Division layout.

4-11

Railroads use trucks to inspect rails and transport heavy tools and parts. Rich Cobb kitbashed this Boston & Maine rail maintenance truck using a variety of parts.

5-1

CHAPTER FIVE

Locomotives of the 1950s

By Robert S. McGonigal

A scene on the Great Northern at Fergus Falls, Minn., illustrates the diversity of the postwar motive-power scene as an Electro-Motive NW3 on a local passenger train meets a 2-8-2 on a freight. Linn H. Westcott photo

The most profound and visible change in North American railroading – the shift from steam to diesel-electric motive power – was completed in the 1950s, **5-1**. At the beginning of the decade, nearly two-thirds of the locomotives on the Class 1 railroads (the hundred-odd major carriers that comprised more than 95 percent of the industry) were powered by steam, while slightly more than one-third were diesels. Ten years later, 97 percent of the railroads' locomotives were diesels, and the handful of steamers still on the books were in storage, never to run again. This dramatic transition, when the last of the great steam engines shared the rails with a profusion of striking new diesels, **5-2**, made the decade one of enduring fascination for railfans and model railroaders.

Virtually from its inception around 1830, the industry had relied on steam locomotives. Electrics first appeared in 1895 but never accounted for more than 2 percent of the total fleet. Diesels emerged in the mid-1920s as switchers, began to be used on high-speed passenger trains in the mid-1930s, and by 1940 proved they could handle main-line freight trains, the most common and demanding work in railroading.

World War II retarded the spread of diesel power, but then the floodgates opened. By the end of the 1940s, nearly every railroad had committed to replacing its steam fleet with diesels, and a few had completed the job. Why did they do it?

Diesel-electrics, which use an oil-fueled engine to turn a generator that feeds electricity to axle-mounted traction motors, are far more efficient than steam locomotives in converting the energy contained in their fuel to useful work. Steam locomotives, which burned coal (or, less commonly, oil) to convert water into steam, had to stop often during their runs to take on water. Diesels had essentially no need to stop for water and required refueling less often than steam. Diesels were much easier to service and maintain than steam engines, resulting in lower costs, higher availability, and longer runs between servicing stops.

Tradition and the nature of the two technologies favored customization of steam designs and mass production of diesels, meaning greater interchangeability of parts for the latter. Unlike steam, a diesel has its maximum power available at starting, allowing dieselized railroads to run longer trains and eliminate helper districts. With their shorter rigid wheelbase and lower axle loadings, diesels were easier on track and more flexible than steam engines.

Carriers in territory where water and/or fuel were not readily available tended to go for diesels early. Some coal-hauling roads, loathe to alienate their shippers, stuck with steam after others had decided to dieselize. The last domino to fall was the Norfolk & Western Ry., which didn't get its first diesel until 1955. The pace of a given road's dieselization depended on the carrier's size, financial status, age and condition of its steam fleet, traffic patterns, and other factors.

Steam locomotives

Steam locomotives of the 1950s were nearly as varied as the railroads that operated them, the builders that produced them, and the services in which they worked.

In general, each of the major railroads, working with a commercial

5-2

Although EMD's E units outsold Alco's PA and PB passenger diesels nearly 4 to 1, the latter's distinctive aesthetics, here displayed by Atchison, Topeka & Santa Fe no. 67, earned them a devoted following. Southern Pacific, Santa Fe, Missouri Pacific, and New Haven had the most. Thomas L. Taylor photo

5-3

Pacifics on commuter trains stand a few tracks away from a Fairbanks-Morse Erie-built diesel and an EMD E unit just outside Chicago & North Western's Windy City terminal in June 1953. Wallace W. Abbey photo

5-4

Even into the 1950s, the Nickel Plate Road believed its superb 2-8-4 Berkshires to be better than diesels for its fast freight service. Because it dieselized late, NKP had no freight cab units, only road switchers. Don Wood photo

builder or its own shops, designed its own locomotives, resulting in a wide variety of designs. Some roads, like Union Pacific and Chesapeake & Ohio, favored articulated locomotives; others, like Santa Fe and the Erie, stuck with two-cylinder designs. Carriers with lots of marginal track, like Canadian National, went for lightweight engines, while those with robust physical plants, like Duluth, Missabe & Iron Range, specified heavy power.

Most steam locomotives in service in the 1950s were built by one of three companies. The Baldwin Locomotive Works of Philadelphia, Pa., was the largest and oldest builder, dating to 1831. The American Locomotive Co. (Alco) of Schenectady, N.Y., was formed in 1901 from several older firms. The Lima Locomotive Works of Lima, Ohio, began building geared logging engines in the 1870s; although a distant third in production totals, Lima was influential, thanks to its "Super-Power" concepts of the 1920s.

Baldwin and Lima built their last steam locomotives for domestic use in 1949, and Alco in 1948. Also at this time, non-export steam production came to an end at Canada's two major builders: Montreal Locomotive Works and Canadian Locomotive Co. With varied success, MLW, CLC, and the Big Three U.S. steam builders tried their hand at diesel construction.

Some roads built their own steam locomotives. One was N&W and another was the Pennsylvania RR, which until 1946 built thousands of engines at its Altoona, Pa., shops. While the PRR bought from the builders even as it produced its own locomotives, the N&W relied on its Roanoke, Va., shops for all its new power from 1928 on. Indeed, Roanoke was the very last builder of steam for domestic use and the only one active into the 1950s. The last reciprocating steam locomotive built for a common-carrier U.S. railroad, N&W 0-8-0 no. 244, rolled out of Roanoke in December 1953.

In 1950, although diesels had taken over many assignments, steam could be found in nearly all types of service in nearly every region of the country, **5-3**. As had been the case for much of the steam era, different types of trains were given locomotives with different wheel arrangements. In yards, 0-6-0s and 0-8-0s switched cars. Local or branchline freights rated 2-8-0s and light 2-8-2s. A local passenger train might have a 4-6-0, 4-6-2, or light 4-8-2. Depending on the territory (flat or heavily graded), merchandise freights were most commonly powered by heavy 2-8-2s, 2-8-4s, 2-10-4s, 4-8-2s, 4-8-4s, or 4-6-6-4s, **5-4**. Heavy coal and ore trains typically got 2-10-2s, 2-10-4s, 2-8-8-2s, and 2-8-8-4s. On the few top-notch passenger trains that hadn't yet been dieselized, 4-6-2s, 4-6-4s, and 4-8-4s did the honors. Short lines carried on with small 2-6-0s, 2-8-0s, and 4-6-0s, many of which were hand-me-downs from Class 1 roads. The most common wheel arrangement was the 2-8-0.

There were exceptions. Union Pacific powered many freights with two wheel arrangements found on no other road:

Locomotive ownership, U.S. Class 1 railroads

	Steam	Diesel	Other	Total
1945	38,853	3,816	861	43,530
1950	25,640	14,047	807	40,494
1955	5,982	24,786	661	31,429
1960	261	28,278	541	29,080

5-5

While most steam and early diesel locomotives were built for specific jobs, the road switcher was suited for almost any duty. Electro-Motive Division's best-selling general purpose (Geep) line was aptly named. Jim Shaughnessy photo

the 4-12-2 and the famous 4-8-8-4 Big Boy. Delaware & Hudson had immense 2-8-0s that could out-pull its 4-6-6-4s. Pennsy and Western Maryland used large, relatively modern 2-10-0s on mainline freights. Southern Pacific's unique 4-8-8-2 Cab-Forwards handled freights and some passenger trains, while old 2-6-0s worked SP's lesser jobs. N&W relied on the 2-6-6-4 to power its fast freights. Canadian Pacific's last new steam engines were 2-10-4s for Rocky Mountain passenger service.

As the decade progressed, steam retreated. Older, smaller engines were natural candidates for retirement, and many did go to scrap, put out of work by diesels or newer steamers, themselves displaced from their intended service. Proud New York Central Hudsons, for example, ambled through their last days on short locals after diesels bumped them from the top passenger runs.

Sometimes the newest steam was the first to go. Many of the final steam designs were too large to be used on anything but big mainline trains, too complex for maintenance forces trying to juggle conventional steamers and new diesels, or simply too troublesome. The Pennsylvania's duplex engines of the 1940s were all three – and barely made it into the '50s. In 1955, two years after the last Wabash 4-8-4 went cold, a pair of ancient 2-6-0s still worked a branch with a weight-restricted bridge.

Diesel locomotives

Diesel locomotives of the 1950s were far more standardized than steam power. There was still plenty of variety, however, as there were six major diesel builders, which together catalogued dozens of models.

Diesels of the era came in three basic configurations: switcher, cab or carbody (later called covered wagon), and road switcher. The switcher layout featured an operator's cab at one end and a long, narrow hood enclosing the diesel engine, main generator, radiators, and other equipment. Such units had two two-axle trucks, with traction motors powering all axles (giving them a B-B wheel arrangement). Generally operated singly, switchers ranged in horsepower from 600 to 1,200 and were produced throughout the 1950s.

The cab or carbody type was introduced in the mid-1930s. The carbody extended the full width and length of the locomotive. At the front, a lowered nose gave the crew forward visibility and collision protection. With a few exceptions, running gear consisted of two two-axle trucks or two three-axle trucks with an unpowered middle axle (A1A-A1A wheel arrangement).

Dates of complete dieselization

Atchison, Topeka & Santa Fe	1957
Boston & Maine	1956
Baltimore & Ohio	1958
Canadian National	1960
Canadian Pacific	1960
Chicago, Burlington & Quincy	1959
Duluth, Missabe & Iron Range	1960
Gulf, Mobile & Ohio	1949
Illinois Central	1960
Nickel Plate Road	1958
Norfolk & Western	1960
New Haven	1952
New York Central	1957
Northern Pacific	1958
Pennsylvania	1957
Southern Pacific	1957
Southern Ry.	1953
Union Pacific	1959

This Alco S-2 switcher, built in 1943, was part of the company's popular S series. Safety stripes along the frame and on the ends enhance the visibility of the otherwise-black SP 1350. Many roads used similar liveries on their yard engines. SP photo

Generally, four-axle cab units were for freight service and the six-axle ones for passenger work, but many four-axle cab units were used in passenger service. Most such diesels were meant to be operated in sets of two or more by multiple-unit control from the leading cab, or A, unit.

Booster, or B, units had no cabs and were combined with A units in A-B-A or A-B-B-A formation, depending on the amount of power desired. Individual units ranged from 1,200 to 2,400 h.p. Cab-unit popularity, especially for freight work, crested about 1950, with the last freight covered wagons being built in 1957 (passenger cab production ended in 1963).

The road switcher caught on in a big way in the late 1940s, **5-5**. These units resembled large switchers with an additional short hood next to the cab. This configuration possessed excellent sight lines from the cab and was easy to board and dismount during switching operations. Doors on the long hood made components accessible for maintenance. The road switcher was suitable for nearly every type of service – even passenger trains, provided the unit was fitted with certain options (like a steam generator for train heating) and the owner didn't mind its decidedly non-streamlined appearance. Some railroads ran their road switchers long-hood forward, while others designated the short hood as "front." (Low short-hood "noses" were introduced in 1959.) By the mid-1950s, the road switcher had become the standard locomotive design and would remain so for more than three decades. Road switchers of the 1950s had B-B or C-C (three powered axles on each of two trucks) wheel arrangements and packed anywhere from 1,000 to 2,400 h.p.

Diesel builders

Six major diesel builders served the railroads' appetite for new power in the 1950s. Biggest by far was the Electro-Motive Division (EMD) of General Motors, which constructed more diesels than all the others combined. EMD's main plant was at La Grange, Ill., near Chicago; a smaller plant in Cleveland built switchers and road switchers from 1948 to 1954, and a plant at London, Ont., began building diesels in 1950.

Landmark EMD products – all of which were in service in the 1950s – included the *Zephyr* trains of the Burlington Route (introduced in 1934), E-series six-axle passenger cab units (1937), F-series four-axle freight and passenger cabs (1939), GP-series four-axle road switchers (Geeps, 1949), SD-series six-axle road switchers (1951), and SW- and NW-series switchers (1939). Very few railroads were without EMDs during the 1950s.

Alco, the No. 2 steam builder, ran a distant second in the diesel era. The firm partnered with General Electric under the Alco-GE banner from 1940 to 1953 and formally changed its name from American Locomotive Co. to Alco Products in 1956. Plants were in Schenectady, N.Y., and Montreal, Que. Alco's top sellers were the RS-series four-axle road switcher (introduced in 1941), FA and FB four-axle freight cab units (1946), PA and PB six-axle passenger cab units (1946), and S-series switchers (1940), **5-6**.

The third-most-prolific diesel builder of the era was Baldwin, which converted its plant at Eddystone, Pa., (near Philadelphia) from steam production. Baldwin began building switchers in the late 1930s, sold its first road units in 1945, and left the business in 1956. Its biggest sellers were switchers: the VO1000, DS-4-4-10, and S-12. Baldwin also had a line of road switchers in the DRS and AS families. Baldwin's most distinctive products were its cab units, whose body stylings and wheel arrangements inspired nicknames like Sharknose, Babyface, and Centipede.

Fairbanks-Morse (FM) began making diesel switchers at its Beloit, Wis., plant in 1944. Its tall opposed-piston engine necessitated an unusually high hood, giving its H-10 and H-12 switchers a distinctive look. Fairbanks-Morse road switchers ranged in size up to the 2,400

5-7

Several electric districts were converted to diesel operation in the 1950s, but not the Pennsy's New York-Washington-Harrisburg lines. P5a 4737 is at Overbrook, Pa., in 1958. Aaron G. Fryer photo

h.p., six-axle H-24-66 Train Master, the most powerful single-engine diesel when introduced in 1953.

While it was gearing up its Beloit facility, FM contracted with GE's Erie, Pa., plant to construct FM's first cab units, the aptly named Erie-builts of 1945. In 1950, FM launched its Consolidation Line of Beloit-built cab units, but the C-Line was quickly swamped by the growing preference for road switchers. Fairbanks-Morse left the domestic market in 1958. Canadian Locomotive Co. built FM locomotives at Kingston, Ont.

Last to enter and first to leave the diesel market was Lima, which kept faith in steam longer than the other commercial builders. Lima-Hamilton (as it was known after a 1947 merger) built a mere 174 switchers and road switchers between 1949 and 1951. The Lima line was discontinued after the firm merged with Baldwin.

Although today it's the top builder, General Electric had a behind-the-scenes presence in postwar diesels. General Electric supplied electrical components to other builders but did not offer a complete line of locomotives. Its 44-ton center-cab and 70-ton end-cab switchers went primarily to short lines, though a few Class 1s bought them too. Having tested cab units during the 1950s, GE announced it was entering the road-unit market with the U25B in 1960.

While steam locomotives were nearly always black, diesels were at their colorful best in the early 1950s. Roads like Great Northern, Western Pacific, Rock Island, Milwaukee, Atlantic Coast Line, and Minneapolis & St. Louis dressed their diesels in eye-catching, multihued paint schemes. Many roads had different liveries for passenger and freight power and sometimes a third for switchers. Toward the end of the decade, however, as railroads became less concerned with public image and more preoccupied with economy, the fancy paint schemes gave way to simpler, cheaper-to-apply versions.

Electric and other

Electric motive power occupied a tiny but high-profile niche in the '50s. The Pennsylvania, New Haven, New York Central, Long Island, and Lackawanna railroads had extensive "juice" operations in and around New York City. The NH and PRR lines extended north to New Haven, Conn., south to Washington, D.C., and west to Harrisburg, Pa. Out West, the Milwaukee Road had two long segments under wire on its line to Puget Sound. Other roads had relatively short electric operations centered on mountain grades, tunnels, or commuter districts.

Many early boxcab electrics were still on the job in the '50s. Streamlined designs like Pennsy's GG1 and New Haven's EP-4 emerged in the 1930s. After World War II, a few carbody designs like the Little Joe were built; the few electrics built after the mid-'50s were of the road switcher style.

Electrified mileage peaked in 1938; its decline thereafter was caused largely by the diesel, which offered many of the advantages of electrification at less cost and with greater flexibility. Eight steam-railroad electric operations were shut down in the 15 years after World War II, but the big ones on the Pennsy, New Haven, and Milwaukee carried on, **5-7**.

Gas-turbine-electric locomotives seemed poised to challenge diesel-electrics in the early postwar years. The Union Pacific bought a fleet for heavy freight service, but no other railroads acquired any.

With steam, diesel, electric, and even turbine locomotives on the nation's rails, the 1950s were a time of unparalleled motive-power variety. Brightly painted passenger diesels rubbed elbows with ancient steam teakettles. Majestic 4-8-4s passed yards worked by diesel switchers or paused to have an F-unit helper added for a grade ahead. Commuters rode to work behind a 4-6-2 one day and an RS-3 the next. Little wonder, then, that modelers have long looked to this unique era for inspiration.

CHAPTER SIX

Weathering locomotives

By John Pryke

6-1

As the diesel became the locomotive of choice, aging steam engines ran out their final miles without new paint and increasing amounts of visible rust. Dave Frary photo

Even though the diesel-electric locomotive had come of age during the previous decade, according to *Trains* magazine, by 1950 American railroads still had 53.5 percent of their tonnage powered by the venerable steam locomotive. By 1959, however, very few steam engines were left, and those still running showed heavy wear and tear that resulted from little or no maintenance, **6-1**. Many of them could have been classified as rolling "rust buckets."

In 1950, road diesels were predominately cab units, along with some hood units that operated primarily as switchers. These early diesels were often painted in bright colors and kept clean to help foster images of a modern, vital railroad, **6-2**. As the steamers ended up in scrap lines, the diesel rapidly became the workhorse of American railroads, and those once-vivid paint schemes were quickly covered with dirt and grime.

During the latter half of the decade, railroads began to adopt new color schemes for their growing fleet of diesels. In many cases, the new color scheme was simplified, with dark col-

Typical diesel locomotive weathering pattern

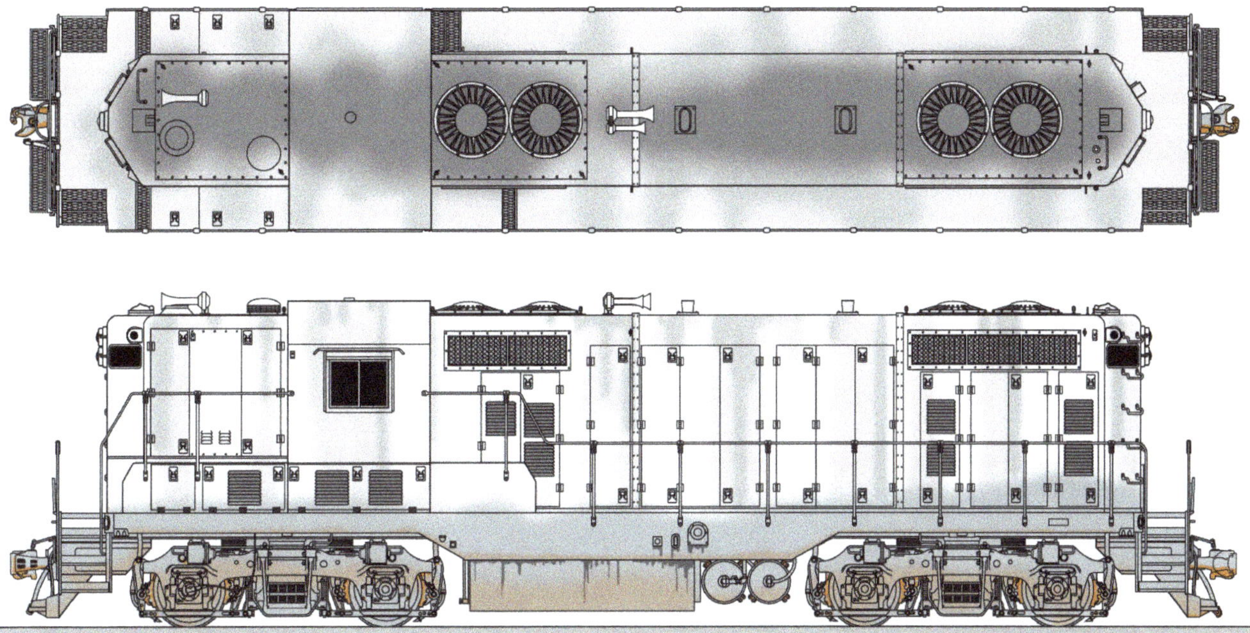

- **Dirt** – truck frames, wheels, fuel tank, underside of air reservoirs, front and rear pilots
- **Ballast dust** – trucks, wheels, lower body
- **Rust** – brake shoes, springs, equalizers, other moving parts and areas with chipped paint
- **Oily black** – roof and cab, streaks down sides of hood or carbody, various fluid fillers
- **Grimy black** – fan and radiator louvers

ors and little striping to minimize the effect of dirt. Some railroads simplified diesel color schemes, so they didn't show the effects of weathering as easily, while others repainted their diesels in color schemes to promote new trains, routes, or services. Consequently, it's perfectly normal to run units with different paint schemes together when you model a 1950s consist.

Today, model railroaders can easily select era-appropriate locomotives (both steam and diesel) from a wide variety of manufacturers. Making these engines appear as though they're operating in the 1950s is primarily a matter of how you apply a variety of weathering effects. Here are some of the techniques I use.

6-2

This New Haven Alco B unit features a vivid red, white, and black paint scheme from the mid-1950s. Photos 6-2 through 6-13 by the author

Running gear wash colors – by region

- **Northeast – brown earth**
- **Southeast – red clay**
- **Midwest – tan earth**
- **Mountain States – gray earth**
- **Desert/Far West – mud/dust**
- **Northwest – brown earth**

Weathering diesel locomotives of the 1950s

While the models we buy are typically sold in new, unblemished condition,

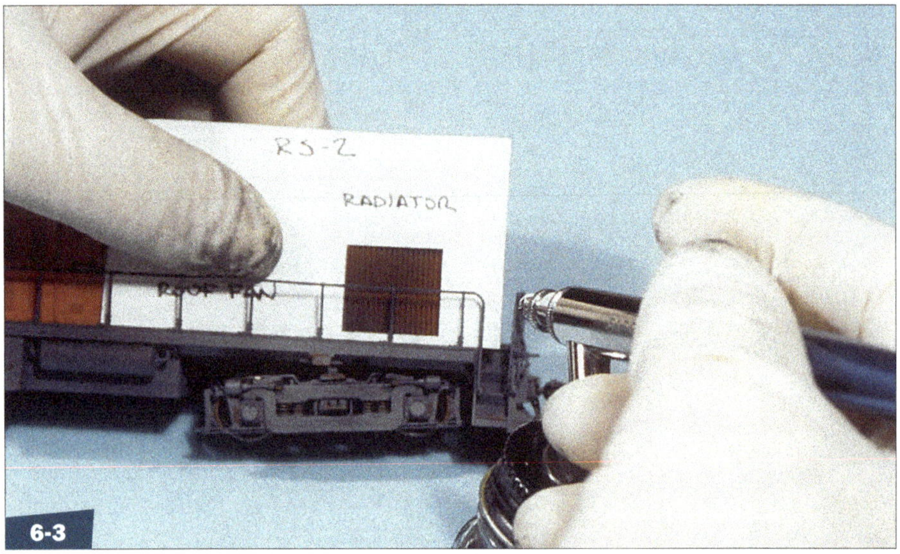

6-3

When spray-painting radiator louvers with acrylic paint, a stencil protects the surrounding area from overspray, and latex gloves keep your hands free of paint.

6-4

The darkened louvers will look as though they've been covered with dirt and grime.

6-5

Spraying Oily Black paint over the top of the hood and center of the cab gives the appearance of exhaust residue.

real locomotives rarely stay that way for long. From the minute its prime mover first comes to life, a diesel locomotive will begin to show the effects of operation and exposure to natural elements.

RADIATORS AND FANS

Radiator and fan louvers turn black from the road dirt that passes through them. You can replicate this effect by cutting a rectangular opening that's the same size as the louvers in an index card. Lay the resulting stencil on the louver to protect the surrounding area, **6-3**. Now spray the louver with Grimy Black paint – a light coat for a new engine and a heavy coat for an older one. When you remove the stencil, the louvers will look as though they've been darkened by dirt and grime, **6-4**.

EXHAUST STAINS

The exhaust from a diesel engine is an oily smoke that tends to settle on the roof, **6-5**. Use an airbrush to spray Oily Black paint over the top of the hood and center of the cab to simulate the accumulated exhaust residue.

If a diesel switcher idles for a long time, the oily deposit will often run down the side of its hood, resulting in a black vertical streak, **6-6**. You can re-create this effect using Oily Black paint applied with an airbrush in vertical strokes.

TRUCKS AND LOWER CARBODY

The longer a diesel has been in service, the more dirt and dust accumulates on its running gear, **6-7**. The color of this deposit tends to match the geography where the diesel operates, so a quick reference table (shown on page 33) helps me determine the most appropriate shades to apply.

When it comes to applying the colors to the trucks, I first attach electrical clip leads to power the motor. Keeping the wheels in motion while you paint prevents a "shadowing" effect from appearing on the wheel faces. I spray the turning wheels and sideframes with Grimy Black, followed by a light spray of dirt and ballast dust colors over the trucks and lower carbody.

A road diesel hauling fast mainline freights raises enough dirt and dust that it also covers most of its carbody. Finally, I highlight the moving parts of the trucks, including springs and brake shoes, with a rust color that I apply using a no. 5/0 brush, **6-8**.

CHIPPED PAINT

The most common place for chipped paint to appear is on the engine's steps – particularly under the cab where the crew's boots kick the carbody, the striker plate above the pilot, and the sheet metal on the pilot and nose, **6-9**. To simulate freshly chipped paint, dip a stiff-bristle brush into Flat Aluminum paint and blot it nearly dry on a paper napkin. Then, very gently press the ends of the bristles against the carbody, leaving tiny aluminum spots that look like the bare metal under the paint. Be sure to blot the bristles almost dry or else you'll end up with blobs rather than dots of paint.

Weathering steam locomotives of the 1950s

Even at the height of the Steam Age, the average steam locomotive never looked clean. Coal dust, road dirt, water streaks from cleanouts, and rust on pipe joints made well-maintained locomotives look dirty and weathered. While not all steam engines weather in exactly the same places or at the same rate, they do show age and wear in a predictable pattern, **6-10**. (For more, see my article, "Realistic steam locomotive weathering," in the September 2002 issue of *Model Railroader*.)

I apply most of the weathering on a steam model with an airbrush, although I paint small details with a no. 5/0 paintbrush. With my Badger no. 150 dual-action airbrush, I use a medium tip for the base coat and a fine tip for details.

CHANGING COLORS

When new, locomotives were painted with shiny black enamel, but in operation, each component turned a different color. All paint became a grayish-black color from smoke, soot, and the weather, and its shine disappeared.

Water leaks and condensing steam from washout plugs, pop valves, and whistles landed on the boiler, leaving light gray streaks and deposits of scale.

When adding vertical exhaust deposits, start at the top and make the streaks lighter as you move down the side.

Roadbed dirt extends up from the trucks and onto the sides of this diesel, which also shows additional light streaks of grime.

Springs, brake shoes, and other moving parts on trucks can be highlighted with rust-colored paint.

6-9

On diesels such as this New Haven DL-109, chipped paint appears most often on the engine's steps, the striker plate above the pilot, and the sheet metal on the pilot and nose.

6-10

In 1953, this aging Mogul shows overall grime and rust resulting from a lack of upkeep.

Typical steam locomotive and tender weathering patterns

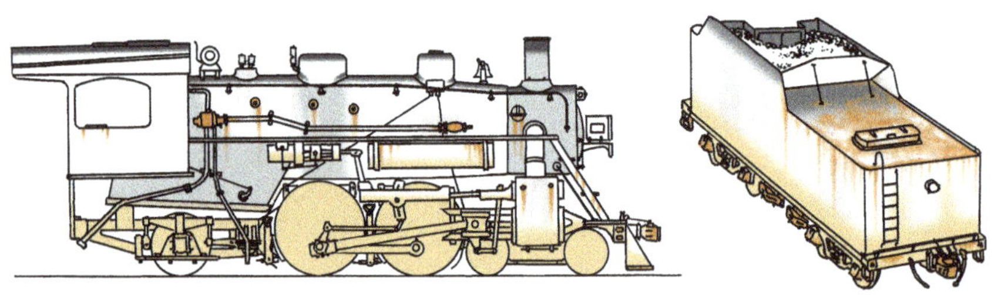

- **Ash gray** – smokebox, firebox, boiler above firebox and below pop (safety) valves and whistle
- **Light gray** – boiler scale drips under washout plugs, under pop valves and whistle, and around check valves
- **Rust** – pipe joints, washout plugs, brake shoe faces, couplers, fittings, springs, and journal box lids
- **Dirt** – frame, wheels, lower edges of pilot and cylinders, underside of air reservoirs, truck frames
- **Soot** – over top of engine, from smokestack
- **Rust** – blotches, drips, glad hands, coupler, truck springs
- **Dirt** – trucks, lower body, splash streaks on rear
- **Coal dust** – deck under slope sheet, top of coal boards
- **Light gray** – streaks on side from condensed steam

The vibration of drivers on the rail kicked up dirt from the roadbed, coating the running gear, trucks, pilot, lower part of the tender, and rear tender beam. The color of this dirt depended on where the locomotive ran.

Parts subjected to constant wear and rubbing – coupler faces, air hose glad hands, polling pockets, and such – rapidly lost their paint and turned rusty from water hitting the bare metal.

JUST ADD RUST

During the 1950s, aging steamers got the minimum maintenance necessary to keep them running. No new paint was applied, and rust rapidly appeared on many parts of the locomotive. You can create a steam locomotive of the

1950s by first applying overall weathering effects and then adding rust to specific areas on the locomotive and tender.

At the front of most steam engines, you can apply rust paint to the coupler knuckle, the air hose glad hands and air valves, the pilot, pilot deck, and steps, **6-11**.

Most water and steam fittings, such as those on the injector under the cab, developed leaks as the packing aged and rust quickly developed on the pipe joints. Trailing truck springs, journal lids, and firebox fittings are also areas where rust typically appeared on the locomotive, **6-12**.

WEATHERING THE TENDER

The tender had its own unique weathering pattern. When water was taken on, spills sat in puddles on the tender deck and filler hatch, settled into the crease between the deck and side, and dripped down the side, often following vertical rivet lines. Rust formed as paint wore off and the water dried. Truck springs, journal lids, the ends of poles, rerail frogs, and the steps are other areas that quickly rusted, **6-13**.

Also on the sides of the tender, broad, light gray streaks often appeared from condensing steam and left a deposit of scale.

Dirt kicked up by the engine coated the trucks and lower sides of the body, and coal dust settled on the deck.

6-11

For best results, use a no. 5/0 fine-tip brush to apply rust to areas on the front of a steam locomotive.

6-12

Rust quickly appeared on pipe joints, truck springs, journal lids, and firebox fittings.

6-13

On the tender, rust can be applied to springs, journal lids, pole ends, rerail frogs, and the steps.

7-1

CHAPTER SEVEN

Passenger equipment and operations

By Robert S. McGonigal

Like many trains of the era, Texas & Pacific's *Westerner*, seen at El Paso in about 1953, carried a mix of older heavyweight and newer lightweight cars. Linn H. Westcott photo

The railroad industry entered the 1950s with a wavering optimism about its passenger service. Fleets of fine new trains crisscrossed the country, but ridership was not living up to expectations, and costs were soaring. One by one, the railroads concluded that passenger trains were a money-losing distraction from their primary business, the movement of freight. By the end of the decade, although there were still a few roads committed to carrying people, the overall picture was dark, prompting *Trains* magazine to devote an entire issue (April 1959) to the question: "Who Shot the Passenger Train?" Whatever their contribution to the balance sheet, passenger trains of the 1950s were colorful, important, and fascinating creations.

In 1900, few highways existed between cities, and railroads dominated travel. Automobiles began to siphon off local traffic after World War I. The decline steepened dramatically and spread to medium- and long-distance trains during the early years of the Great Depression. Streamliners – the lighter, faster, more comfortable, often diesel-powered trains introduced in 1934 – put a bright new face on rail travel and slowed the slide in ridership.

Then came World War II, which saw passenger traffic jump to more than four times its prewar levels, making that segment of the railroad business more profitable than it had been in decades. But wartime production restrictions limited the carriers' ability to acquire new rolling stock, which meant that in 1945 the passenger car fleet was old and nearly worn out.

Thus the stage was set for an unprecedented upgrading of the nation's passenger trains. Even before the end of hostilities in Europe, the railroads began placing orders for a new generation of streamlined cars, and soon they began flooding the rails – more than 3,200 were delivered between 1946 and 1949. The peak year was 1950, when 1,078 new cars arrived. After a lull, deliveries averaged 371 a year for the 1953-56 period but then sank for good – only 66 cars entered service in 1959, the lowest total for a peacetime year since 1933.

7-2

The dome car was the most spectacular innovation in the railroads' push for passenger business after World War II. In a rearward view in the dome of a Canadian Pacific car, a woman ascends the stairs from the car's main level. Budd Co. photo

PASSENGER TRAINS

A cornucopia of cars

All of this new streamlined equipment allowed the retirement of thousands of heavyweight cars dating from the 1920s, although many worked alongside the lightweight newcomers (as well as many prewar streamliners) into the 1960s, **7-1**. Hundreds of old-timers got a new lease on life through rebuilding programs that included air-conditioning, updated interiors, and wide picture windows. Sometimes, higher rooflines and six-wheel trucks were the only external signs that a car was a rebuilt heavyweight and not a new lightweight.

The car fleet of the 1950s included long-established types like coaches, sleepers, diners, lounges, observation cars, baggage cars, and Railway Post Offices. Following a trend begun in the late 1930s, most new sleeping cars dispensed with the traditional open-section accommodation (two seats facing each other, which converted into upper and lower berths separated from the corridor only by curtains) in favor of individual rooms of various sizes and configurations. The 12-1 sleeper (12 sections, 1 drawing room), workhorse of the heavyweight era, gave way during the late 1940s and '50s to the 10-6 (10 roomettes, 6 double bedrooms) and other all-room types. In 1956, the Slumbercoach was introduced; identifiable by their staggered windows (also used on previous duplex roomette cars), these cars offered private sleeping quarters at budget fares.

The landmark car type of the postwar era, however, was the dome car, **7-2**. Conceived in 1944, the dome, which allowed passengers in an upper-level, glass-enclosed seating area to see in all directions, had its first big public exposure in 1947, when General Motors' *Train of Tomorrow* began a nationwide tour and Chicago, Burlington & Quincy's *Twin Cities Zephyr* got new equipment. The dome cars on both trains were immensely popular; ultimately, 17 railroads, mostly in the West, ordered them. Most domes occupied roughly the middle third of the car, but three railroads (Santa Fe, Milwaukee Road, and Great Northern) had full-length domes.

Another car emblematic of the 1950s, the Rail Diesel Car (RDC), also had a humpback appearance. An updated version of the self-propelled "doodlebug" of decades before, the RDC, or Budd car (for its builder, the Budd Co.), incorporated stainless-steel construction, twin underfloor diesel engines (plus rooftop radiators), bidirectional operation, multiple-unit capability, air-conditioning, a modern interior, and five body configurations.

Introduced in 1949, the RDC sold to 28 railroads to the tune of 398 units over 14 years. Rail Diesel Cars were most commonly bought for local or commuter service, but a few roads, including Western Pacific and Baltimore & Ohio, used them on runs of hundreds of miles, **7-3**.

Going up, going down

Unique to the Santa Fe were the double-deck Hi-Level cars delivered in 1956 for the road's Chicago-Los Angeles *El Capitan*. These landmark cars featured side entry doors on the lower level and end doors for passage between cars on the upper level. Santa Fe's Hi-Level coaches could seat up to 72 passengers, versus 44 in standard long-distance coaches. Hi-Levels also came in diner and lounge versions, the latter with partly glass roofs. Santa Fe's Hi-Levels were the template for Amtrak's Superliners of the 1980s.

Commuter-train operators also looked skyward to increase seating capacity. The Burlington introduced the gallery car to its Chicago suburban runs in 1950. These cars featured a conventional seating arrangement on

7-3

The self-propelled Rail Diesel Car, developed by Budd in 1949, enjoyed widespread popularity in a variety of services; this is a Canadian Pacific car in Nova Scotia.
Jim Shaughnessy photo

the main level, above which were two longitudinal balcony-like structures with single seats. Chicago & North Western and Southern Pacific soon followed suit. Also known as bi-levels, these cars could be mixed with single-level rolling stock and were even pulled by steam locomotives at first.

By contrast, the desire for economy prompted experimentation with low-slung, ultra-lightweight trains for medium-distance runs. While they were cheaper to build and operate than standard equipment, their rough riding qualities and spartan amenities failed to impress travelers increasingly turning to automobiles and airliners. Most of the ultra-lightweights had short lives, but two designs, GM's *Aerotrain* and Budd's *Keystone*, lasted into the 1960s on the Rock Island and the Pennsylvania, respectively.

Long-distance trains

The most-glamorous, best-remembered trains of the 1950s plied routes of several hundred miles or more between major cities. The most elite, like New York Central's *20th Century Limited* (New York-Chicago), Santa Fe's *Super Chief* (Chicago-Los Angeles), and Illinois Central's *Panama Limited* (Chicago-New Orleans), carried no coaches, only sleepers, diners, and lounge cars (and perhaps a baggage or mail car).

More numerous were long-haul trains whose consists also included coaches, like Southern Ry.'s *Crescent* (New York-New Orleans), Baltimore & Ohio's *National Limited* (New York-St. Louis), and Northern Pacific's *North Coast Limited* (Chicago-Seattle/Portland). The all-coach, long-haul train, which helped revitalize the passenger business in the 1930s, faded quickly (with some exceptions, like Santa Fe's *El Capitan*) after World War II in the face of automobile competition. Many of these trains operated over more than one railroad during their runs.

More often than not, bringing up the rear of the top trains would be an observation car with a lounge area in the rounded rear end and sleeping rooms in the forward part of the car. Most of the premier cross-country trains received new cars in the immediate postwar years, and such modernization often included dome cars.

Domes usually had coach seating on their main levels, but some had sleeping accommodations, lounge space, or dining areas (Union Pacific even served meals at tables in the dome). Perhaps the ultimate postwar cars were the dome-observation-lounge-sleepers built in 1949 for the Burlington-Rio Grande-Western Pacific *California Zephyr* (Chicago-Oakland) and in 1955 for Canadian Pacific's *Canadian* (Montreal/Toronto-Vancouver).

Lesser long-haul trains often sported a mix of heavy- and lightweight equipment. Running on slower schedules and stopping at towns the flyers passed by, secondary trains tended to have significant numbers of head-end cars for mail and express shipments. Long-haul trains were seldom less than eight cars long and could range up to 15 or more.

Carbuilders of the 1950s

American Car & Foundry (ACF Industries after 1954) built passenger cars in the postwar era at its Berwick, Pa., and St. Charles, Mo., plants. The smallest of the Big Three postwar car-builders, ACF produced some memorable cars, including the Union Pacific domes and the *Mountain*-series observation cars for Great Northern's 1950 *Empire Builder*. In 1959, ACF switched to freight-car construction exclusively.

The Budd Co. built the most dome cars, all the RDCs, and the cars for such famous trains as the *Zephyrs*, the *Canadian*, the *Congressional* and *Senator*, and the Hi-Level *El Capitan*. Nearly all its cars had the stainless-steel fluting that came to be emblematic of the streamliner era. The second-most-prolific builder, Budd produced cars at a plant near Philadelphia.

Pullman-Standard was the carbuilding arm of Pullman, Inc., whose Pullman Co. subsidiary operated sleeping cars for the railroads. (An antitrust suit resulted in the Pullman Co.'s sale to a consortium of railroads in 1947, breaking up the car-building and operating functions.) By far the biggest of the builders, Pullman-Standard supplied cars of all types to nearly every railroad. Its plant was in Chicago.

Regional trains

Five decades ago, many notable trains ran between cities 100 to 500 miles apart, routes that in the Amtrak era would be dubbed "regional corridors." The Chicago-Minneapolis/St. Paul market supported fast trains on three railroads. On the West Coast, Southern Pacific's *Daylight* was the way to go between LA and San Francisco. The Pennsy and New Haven fielded dozens of trains, some 18 cars long, in the 457-mile Boston-New York-Washington corridor.

Running mostly on daytime schedules, regional trains carried coaches, lounges, diners, parlor cars, and observation cars but few if any head-end cars. One of the shorter runs, the Reading Co.'s five-car *Crusader* (Philadelphia-Jersey City), had two observation cars – one on each end – to avoid turning the consist at the end of each trip. Some parlor cars on intercity trains had drawing rooms for travelers desiring privacy. Wabash's *Blue Bird* (Chicago-St. Louis) even had dome cars.

Locals, commuters, and troop trains

Although they were especially vulnerable to highway competition, and so declined in numbers significantly, local and branchline trains were a colorful part of 1950s railroading. Using mostly hand-me-down heavyweight coaches and the oldest motive power, **7-4**, such trains might plod along main lines for hundreds of miles, stopping at every hamlet to exchange mail bags and milk cans.

Others, like Norfolk & Western's "Virginia Creeper" (Bristol, Va.-West Jefferson, N.C., actually a mixed freight-and-passenger train) would wander along remote branches, providing a link to the outside world. Locals could be as short as a single engine and one car or as long as a dozen or more head-end cars and a small handful of coaches.

Odd juxtapositions could occur on locals: a new road switcher pulling ancient cars, a 1920s doodlebug paired with a streamlined postwar coach, or a large, modern steam engine on a short consist.

In the 1950s, as today, the majority of rail passengers used trains to get to work and back on weekdays. Commuter, or suburban, service was most intense in the big cities of the Northeast, in Chicago, and in San Francisco. A few other cities, like St. Louis and Milwaukee, had mainline locals that were primarily commuter runs. Around New York, Philadelphia, and Chicago, where traffic was particularly dense, railroads began electric operation with self-propelled multiple-unit cars in the first third of the 20th century. Most commuter services were still going strong, if unprofitably, in the 1950s.

Because equipment utilization was so low outside the weekday rush hours, some commuter operations were bastions of outdated locomotives and rolling stock. Much equipment, particularly coaches, was designed for commuter work, although downgraded mainline cars weren't uncommon. Train length ranged from a couple of cars on midday or weekend runs to a dozen during rush hours.

For many months after the end of World War II in August 1945, trains carrying troops home were a common sight. Many used the 2,400 boxcar-like troop sleepers and 800 kitchen cars built between 1943 and 1946. The outbreak of the Korean War in June 1950 brought a return of troop trains but without the distinctive sleepers. Sold off by the government in 1947, troop sleepers (and some kitchens) had been converted to boxcars, express and baggage cars, steam-generator cars, and even cabooses. Korea-bound troops had been accommodated in heavyweight sleepers but were often fed from the remaining kitchen cars.

The 1950s was the last decade in which passenger trains really mattered to the railroads and to America. In the years since, trains for the most part haven't been nearly as numerous, varied, or exciting.

Not all trains of the '50s were shiny new streamliners. Many, like this Baltimore & Ohio local, were equipped with locomotives and cars from the 1910s and '20s. Richard J. Cook photo

7-4

CHAPTER EIGHT

Detailing the classic El Capitan

By Buzz Lenander

8-1

A childhood trip on Santa Fe's *El Capitan* and a set of Hallmark imported brass Budd Hi-Level cars inspired me to model this famous train. Photos by the author

Growing up in Albuquerque, I saw many of the Atchison, Topeka & Santa Fe's famous passenger trains, including the *Chief, Super Chief*, and *El Capitan*. Of all these trains, *El Capitan* was my favorite – I actually rode on it in 1956 and have long had an interest in modeling the train, **8-1**.

To make my N scale version of the Santa Fe's 1956 *El Capitan,* I started with two pairs of Kato's Electro-Motive Division F7A and F7B. The Santa Fe had re-equipped *El Capitan* with Budd Hi-Level passenger cars during the summer of 1956, so I used a set of Hallmark imported-brass Budd cars for these. To round out my train I included a baggage car built from a Des Plaines Hobbies kit. Once I'd acquired the necessary locomotives and cars for my *El Capitan*, I set about detailing and decorating the models to match the prototype.

Made in 1982 with just 80 sets offered, the imported-brass Hi-Level Hallmark models may be hard to come by these days, but they do appear occasionally on online auction sites. However, even if you can't find them, you can still use many of these techniques to model other passenger trains.

Detailing Kato F7 A and B engines

I started my *El Capitan* modeling project by detailing the A-B-B-A set of Kato EMD F7s, **8-2**. The models feature nicely painted shells, so I worked carefully to detail them without having to repaint them. I added an assortment of details including metal grab irons, lift rings, diaphragms, rear-view mirrors, handrails, and etched-brass windshield wipers.

Santa Fe F7s had stainless-steel side panels – six per side on the A units and 12 per side on the B units. To create the stainless-steel effect, I used Bare-Metal Ultra Bright Chrome Foil. This metal foil has an adhesive back, and it conforms to most surfaces very well without hiding details.

Before applying the foil, I used a no. 17 chisel blade to remove the inboard handrail on the lower end panel of each unit. I then sanded the surface with a 600-grit sanding stick.

Using a hobby knife with a brand-new blade, I cut the foil for the panels with the backing paper still attached. Then, working on one panel at a time, I peeled off the backing paper and laid it on the shell. I burnished the foil with a soft, flat balsa stick wrapped with a piece of nylon cloth. I trimmed excess foil with a hobby knife.

The two front panels of the A units were the hardest to cut, **8-3**. To help match the contour of the striping, I made several copies of the warbonnet outline from a sheet of Microscale decals and then traced over them to cut the foil pieces.

The B-unit panels are all rectangles, but I had to cut openings for the three portholes and two sand hatches on each side, **8-4**. I cut these pieces with a sharp knife, but only after I'd thoroughly burnished the foil to the surrounding panel.

I used Microscale decal sets 60-48, 60-101, and 60-793 to replace the lettering covered by the foil.

Riding El Capitan

During the summer of 1956, when I was 10, I spent a week with my grandparents in McPherson, Kan. (My grandfather had worked for the Union Pacific for 52 years, so that's how railroading got in my blood.) My parents purchased a ticket for me to travel from Albuquerque to Hutchinson on the Santa Fe's *El Capitan*. At that time, the train had been freshly outfitted with the railroad's new Budd Hi-Level cars.

My round trip ticket cost $17.45, plus an additional special service charge of $1.10 each way for a reserved coach seat. (I still have the ticket coupons.) I departed Albuquerque at 6:55 a.m. on Aug. 21 riding train 22 eastbound. My ticket was for chair car no. 222, seat 37; I returned on train 21 on Aug. 29 riding chair car no. 216. (Car numbers were different on eastward and westward trains. On train 21, the equivalent of car 222 was car 212.)

It's been almost 50 years since I rode the Hi-Level *El Capitan*. Though many things have changed since then, Amtrak's *Southwest Chief* still arrives and departs Albuquerque each day with stops in Hutchinson.

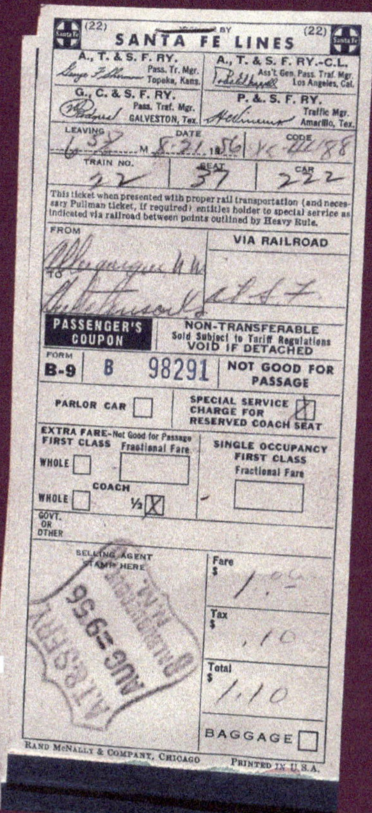

I installed Micro-Trains Z scale couplers on the front of each A unit since they are closer to scale size than traditional N scale knuckle couplers. For the rear of the A units and both ends of the B units, I used Micro-Trains no. 1128 MT-7 couplers with adapters, **8-5**. These couplers move back and forth horizontally, allowing the diaphragms to work better as the engines run through curves.

New lights and other details

Next, I used a no. 70 bit to drill out the classification lights molded on the shell. I installed new lenses made from .020" fiber-optic strands. I formed the lens faces by heating the end of a fiber-optic strand over an open flame for a second. I then inserted the strand in the hole in the shell and cemented it with cyanoacrylate adhesive (CA). Once the CA had dried, I clipped the

F7A F7B F7B

I detailed the front of the A units by adding windshield wipers, handrails, grab irons, and m.u. hoses.

Applying the foil on the front panels of the A units took some trial and error before I was satisfied. I painted the plastic horn castings, handrails, and mirror Polly Scale Soo Line Red to match the locomotive's paint and added the speed recorder.

strand flush with the inside of the shell.

Santa Fe F7As had two headlights. The upper light contained an oscillating signal light and a red emergency lens. By 1956, the lower casing on some F units had two sealed-beam headlights mounted horizontally.

To represent these locomotive lighting effects, I used a Richmond Controls EZ03 Kato F7 lighting kit. After installing the kit, I made new lenses for the headlights from .005" clear sheet styrene and simulated the lightbulbs with fiber-optic strands.

One of the last details I added to the locomotives was a speed recorder on each A unit. I used Sunrise Enterprises speed recorder and cable parts for these, **8-3**.

Santa Fe took pride in its passenger trains and kept them pretty clean. During the stopover in Albuquerque, the train was washed by mobile car washers with rotary brushes that extended to the top of the Hi-Level cars. As a result, I only lightly weathered the locomotives, using powdered pastels. I put a little black around the roof fans and along the grills and a little gray dust on the truck side frames and pilot.

The cars of *El Capitan*

In the early 1950s, railroad passenger traffic patterns were in the midst of some major changes. The average long-distance rail traveler purchased tickets for coach or chair car seating rather than traditional sleeping car service.

Inaugurated in 1938, *El Capitan* was the Santa Fe's premier chair-car train. In 1956, the railroad re-equipped this train with new Budd Hi-Level cars as a cost-saving measure. The increased capacity of the double-deck cars meant the Santa Fe could use fewer cars to carry the same number of passengers.

After delivering two prototype Hi-Level chair cars in 1954, Budd built 47 additional cars for the Santa Fe in 1956. This gave the railroad the five *El Capitan* consists needed to protect the daily schedule of trains 21 and 22. *El Capitan*'s Hi-Level fleet included six diners, six lounges, and 37 chair cars.

On July 8, 1956, two matched sets of Hi-Level *El Capitan* cars left their respective terminals of Chicago and Los Angeles for their inaugural runs. Six weeks later, I rode the Hi-Level *El Capitan* from Albuquerque to Hutchinson, Kan.

TRANSITION CARS

The Santa Fe had six transition cars, nos. 3477 to 3482, but these cars were not all alike. The Hallmark car is modeled after one-of-a-kind car no. 3482. Pullman built this car in 1940, and in 1952, it was converted to a straight baggage-dormitory car. It was 79' 10" long.

Later, car 3482 and five other cars were modified into transition cars for the Hi-Level *El Capitan*. They were all fitted with a streamlined fairing on the vestibule end of the car to cut down on wind resistance at high speeds between the two different-height cars.

F7B F7A Baggage-express (x2)

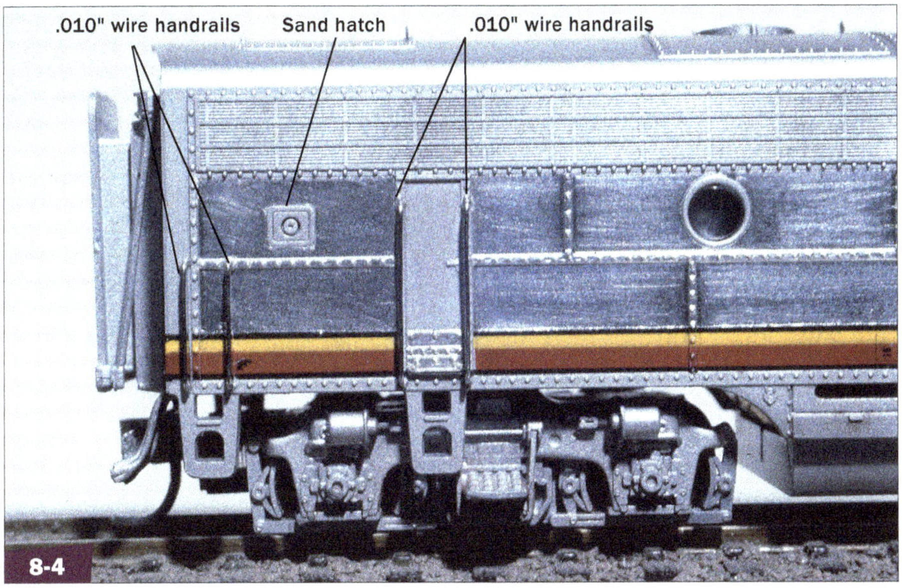

After burnishing the Bare-Metal Ultra-Bright Chrome Foil to the side panels, I cut and removed the foil around the sand hatches and windows. I then replaced the lettering covered by the foil with new decals.

End details I added included Micro-Trains no. 1128 MT-7 couplers, diaphragms, grab irons, and m.u. hoses.

A second modification to these cars was the addition of a radio antenna along the center line of the roof, **8-6**. This was for the train's music system. The antenna is included as an extra part in the Hallmark set but needs to be installed on the model.

Headlight assembly

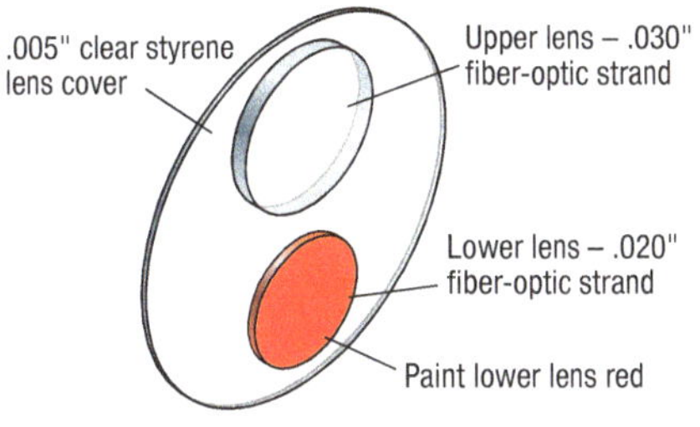

The oscillating headlight with the red emergency light is simulated by cementing fiber-optic pieces to the clear styrene lens.

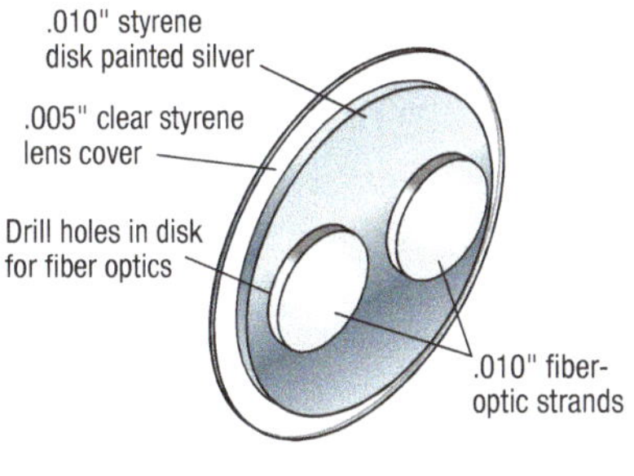

To make the twin sealed-beam lower headlight, cement fiber-optic lenses into holes drilled in a styrene disk. I painted the disk sliver.

CHAIR CARS

Step-down chair cars (528-537 series) were the only cars that provided a connection to conventional-height cars. These cars had 68 seats, all on the upper level. The lower level housed the rest rooms, a baggage area, and mechanical equipment.

There were two step-down cars on each train set – one between the transition car and the Hi-Levels, and one at the rear of the train.

The other chair cars in the train, (700-724 series) were nearly identical to the step-down chair cars in outward appearance. These cars could seat 72 passengers but had high doors at both ends for connection only to other Hi-Level cars.

DINER AND LOUNGE

The diner and lounge cars were in the middle of the train. The diners were numbered 650 to 655. Weighing 97 tons, these cars were the heaviest single-unit diners produced for any railroad. It was the only car in the train to ride on heavy-duty 61-RDO six-wheel trucks. (Other Hi-Level cars rode on 41-CDO four-wheel trucks.)

The diner could seat 80 passengers comfortably, with the upper level almost entirely devoted to seating. The lower-level galley provided the most modern and spacious kitchen of any diner at that time. Dumbwaiters lifted prepared food to the dining room above.

Transition car baggage-dormitory

Step-down chair car

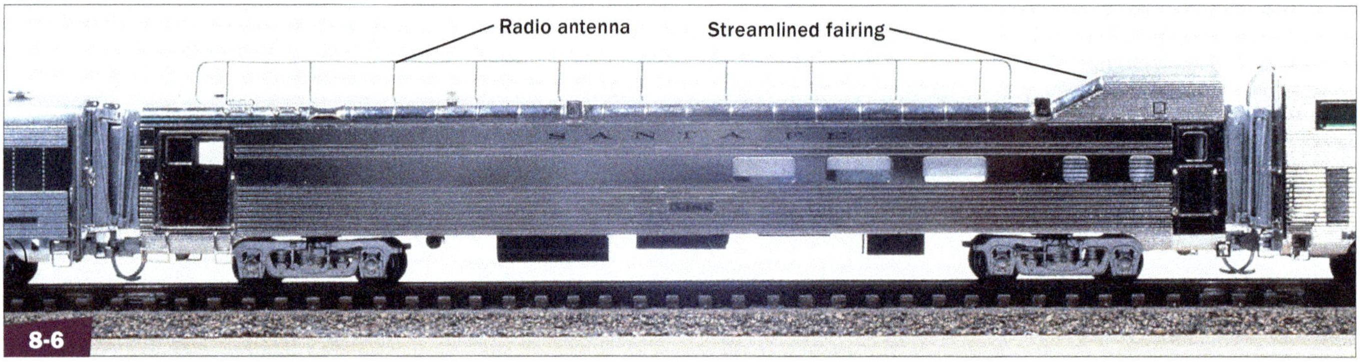

The Santa Fe modified six passenger cars to act as transition cars for use with *El Capitan*, and each was unique. The radio antenna is for the train's music system. Other Santa Fe Hi-Level trains didn't get this feature.

The car assignment numbers appear in the lower part of the men's room window. I cemented the numbers to the shades.

The most impressive car in the train had to be the glass-topped lounge car. The Hi-Level Top of the Cap cars (also called Sky Lounges) were numbered 575-580. Because their larger windows provided a good view and gave the interior a spacious feeling, these cars were an instant hit with the traveling public. The upper level contained a 60-seat bar-lounge area with a newsstand. The Kachina Coffee Shop was a more intimate 26-seat lounge and bar on the lower level.

Hi-Levels in N scale

Fortunately for N-scalers, Hallmark Models commissioned KMT of Japan to build brass models of the Santa Fe Hi-Level cars. Each set contained six cars and included a no. 3482 transition car, two 528-series step-down chair cars, one 650-series diner, one 575-series lounge car, and one 700-series Hi-Level chair car.

The 1956 *El Capitan* consist had 10 Hi-Level cars in it, so I needed four 700-series chair cars to complete my train. A few individual chair cars were available, and the 528-series step-down chair cars could be made to look like the 700-series cars by placing a high-level diaphragm on each end of the car.

Baggage-express cars

The 1956-57 consists called for one storage mail car and one baggage-express car running through between Chicago and LA. Both were lightweight baggage-express cars. Des Plaines Hobbies makes a kit for a 1951 American Car & Foundry baggage-express car, and I built one to use on my train.

The baggage car is a core kit with plated-brass sides. As with the F7s, I used Bare Metal foil to cover the plastic roof and roof vents of the car to match the stainless-steel appearance of the rest of the train. The baggage car model represents one from the 3453- to 3466-series of cars.

Decals

The Hallmark cars are nickel-plated and have grab irons, handrails, and stirrups but no lettering or window glazing. Nor do these cars include any interior or underframe details. (The transition car is the only exception.) Though the remaining cars have a flat bottom, it's not noticeable when the cars are on the track.

I prepared the cars for decals by airbrushing a coat of Polly Scale Clear Gloss on the letter and number boards. Once the gloss coat had dried, I lettered the cars using Microscale set 60-4190. (Microscale set 60-4174 has all the car numbers for *El Capitan*, but the font is too small.)

After all the decals were set and dry, I used Poly Scale Clear Gloss to seal them in place.

Chair car

Chair car

I replaced the original Hallmark trucks (left) with American Limited Models 41-CDO trucks (center). The six-wheel truck used on the diner is the original but with new plastic wheelsets (right).

Glazing, shades, and car numbers

For the head-end cars and the transition car, I added window glazing, using clear .005" sheet styrene. I also used this for all the lower windows in all of the Hi-Level cars. I cemented the styrene in place with CA, applying it to the interior of the car's shell with a Microbrush. Because the brass car sides are thin, the finished window glazing looks nearly flush.

For all of the upper level windows of the Hi-Level cars, I used MonoKote Metallic Green, made by Top Flight. This transparent plastic film is made for covering model airplanes but works great for making tinted windows. With the glazing in place, it's hard to see that my cars don't have any interiors.

In working with the MonoKote film, I found it easiest to cut strips of the material long enough to cover only three or four windows at a time. The MonoKote has a clear coating on one side, so I needed to peel it off before cementing the glazing to the car.

The hardest car to apply glazing to was the Sky Lounge since these windows curve up into the roof. After some experimenting, I found it easiest to use strips that covered two windows.

To further enhance the windows, I added shades. The Santa Fe used aluminum (silver) shades, which I simulated with gray paper. I cut the paper into strips and cemented them inside the carbody. I varied the heights of the shades in the windows, so they didn't look too uniform. I didn't put shades in the head-end cars or on the upper deck of the dome car.

The Santa Fe assigned a car number to each chair car on a train. Passengers were assigned to a specific car and seat. The number was displayed in the small window to the right of the vestibule door on the side of the car with two small windows. It was also in the middle window to the left of the vestibule door on the side with five small windows. These windows were located in the men's restroom on the lower level, **8-7**.

To make the car assignment numbers, I used white numbers set on black backgrounds from Microscale decal set 60-113. I cut the car numbers out of the decal sheet and cemented them to my window-shade paper, using a glue stick. I then glued the shade inside the proper window on the car so that the number appeared at the bottom.

Trucks and couplers

The cast-metal truck frames and wheelsets that the Hallmark cars were supplied with left a lot to be desired, mainly in detail and good rolling qualities, so I replaced them, **8-8**.

For most of the train set, I used American Limited Models 41-CDO four-wheel trucks with outside swing hangers. These trucks are correct for Santa Fe Hi-Level cars. I airbrushed the sideframes silver.

I used 36" plastic wheelsets I had left over from some Con-Cor auto rack cars. These wheels are a larger diameter than those supplied with the trucks, and when properly adjusted, they roll freely.

The diner rode on six-wheel trucks. I replaced the wheelsets but used the existing cast-metal sideframes, so these trucks looked much better than the original trucks, **8-8**.

To attach the new trucks to the car floor, I tapped the bolster holes using a no. 2-56 tap. I then used no. 2-56 round-head screws to hold the trucks in place.

The bottom of each car comes ready to mount a Micro-Trains coupler draft gear. I wanted my cars to couple more closely than the original holes allowed, so I drilled new no. 60 holes just inboard of the original mounting holes, tapping them for 00-90 screws. Next, I airbrushed the couplers and draft gear silver and then mounted them to the cars. The coupler box needs to sit flush with the bottom of the car, so I trimmed off the lip molded on top of the box.

Diaphragms

I added diaphragms to all the cars. For the conventional cars and the step end of the two step-down chair cars, I used

Diner Sky Lounge

Hi-Level diaphrams

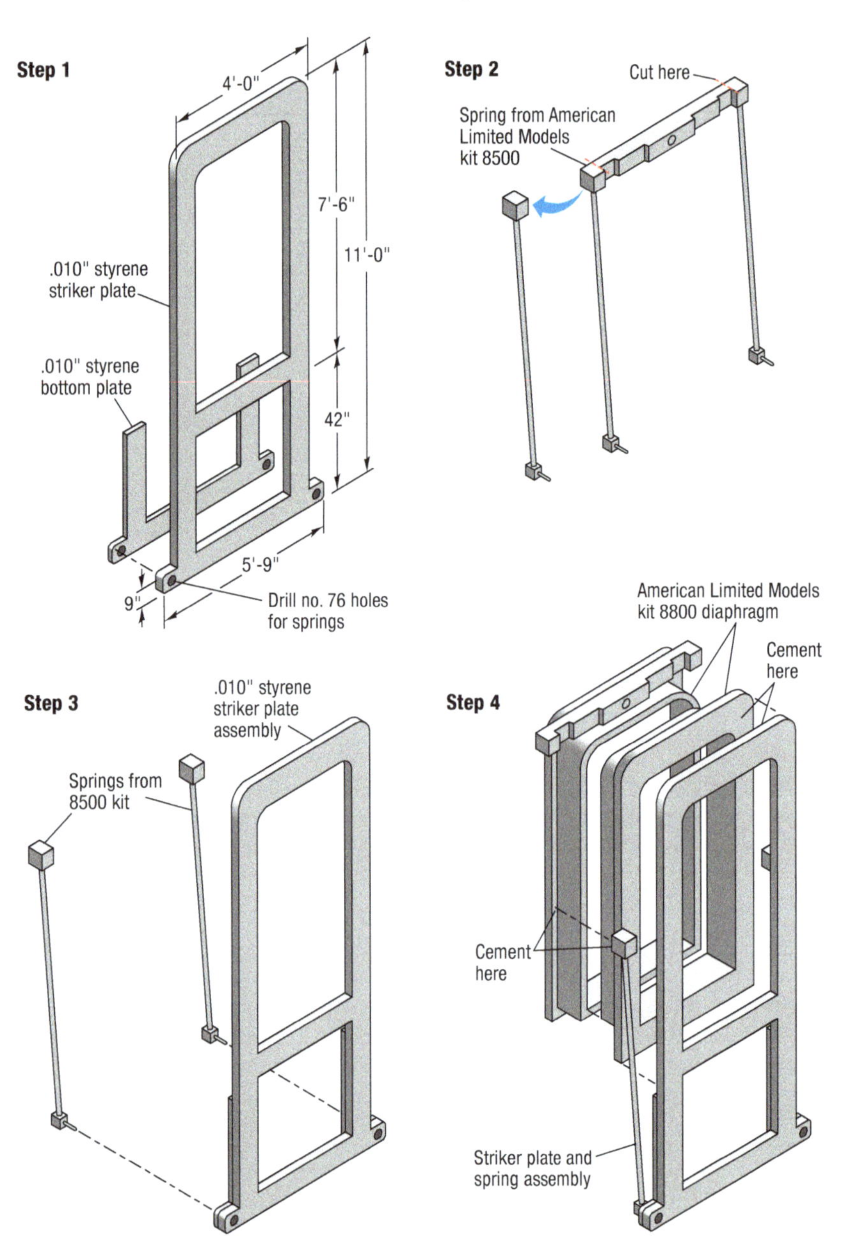

To make the Hi-Level diaphragms, I used parts from two American Limited kits and some sheet styrene.

American Limited Models no. 8500 diaphragms. However, no one offers Hi-Level diaphragms, so I had to make my own. I used American Limited Models no. 8800 diaphragms, which are smaller than those I used on the other cars, and combined them with parts from an 8500 diaphragm kit and some styrene.

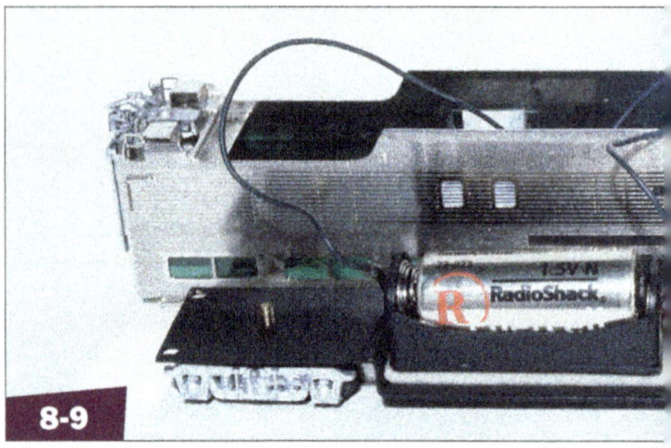
8-9

I started by cutting a new Hi-Level-style striker plate from .010" sheet styrene. As shown in step 1 of the figure at left, the plates are a scale 4 feet wide and 11 feet tall. The new striker plate is actually a copy of the 8800 plate but with an additional 3' 6" extension on the bottom. I cut a second piece of .010" styrene to match the plate extension and glued it to the inside of the Hi-Level plate. I then shaped the bottom of the new plate to match the bottom of the 8500 plate.

I used the plastic springs from an 8500 diaphragm kit, cutting them as shown in step 2. I used a no. 76 bit to drill holes in the bottom corners of the striker plate, and cemented the springs into the holes, as illustrated in step 3. At this point, I assembled an 8800 diaphragm kit and cemented it to the back of the new striker plate, as shown in step 4.

Once I'd finished diaphragms for the entire train, I painted them silver and cemented then to the Hi-Level cars using CA. I also added steam and air lines and uncoupling levers but only to the rear of the tail-end car.

Drumhead

The final detail for the train was a train sign on the end car, **8-10**. I purchased a Tomar Industries AT&SF *El Capitan* Conquistador tail sign kit. On the prototype, a red taillight was mounted above the sign as a marker light, so I cemented a red Miniatronics

Chair car　　　　Chair car

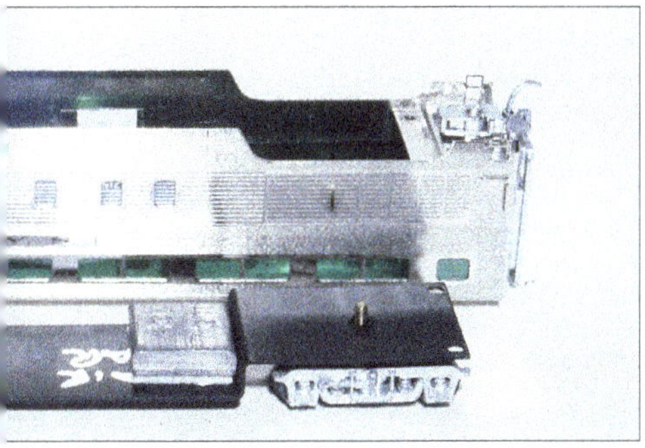

A size N 1.5-volt battery powers the tail sign and marker light. I attached the battery holder to the floor of the last car with double-sided tape.

To match the prototype, I added a red taillight to the top of the drumhead as a marker light.

1.2mm 1.5-volt bulb to the top of the drumhead. I painted all but the front of the bulb silver.

The Tomar tail sign also uses a 1.5-volt bulb, so I powered both bulbs with a size N 1.5-volt battery, **8-9**. I used double-sided foam mounting tape to attach a RadioShack N battery holder to the floor of the car and soldered the leads from the bulbs to the holder. When I run the train, I unscrew the floor, insert the battery, and I have a constant-lighted marker and drumhead.

With the detailing work complete, my *El Capitan* is ready for the rails. "All aboard!"

Materials list

American Limited Models
8100 41-CDO passenger outside swing hanger trucks
8500 passenger car diaphragms
8800 E and F unit diaphragms

Bare-Metal Foil Co.
004 Ultra Bright Chrome foil

BL Hobby Products
.010", .020", and .030" fiber-optic strands

Des Plaines Hobbies
6002 Santa Fe 1951 ACF baggage/express car

Detail Associates
2502 .008" brass wire
3503 .010" stainless-steel wire

Gold Medal Models
160-7 diesel locomotive details
160-26 Kato smooth-side passenger car details

Hallmark Imported Brass
Santa Fe *El Capitan* Hi-Level six-car set
Santa Fe Hi-Level 700-series chair car (4)

Kato
176-2121 Santa Fe F7A (2)
176-2211 Santa Fe F7B (2)

Microscale Decals
60-48 Diesel locomotive data
60-113 MoPac hoods no. 2
60-101 Santa Fe warbonnet passenger diesel cab E and F 1950-70
60-793 E and F units diesel data and number boards 1944 +
60-4190 Santa Fe streamlined general-service passenger cars, 1939-70

Micro-Trains Line Co.
905 Z scale couplers
1023/1025 couplers and draft gear
1128 MT-7 couplers with adapter

Miniatronics Corp.
18-001-10 1.5V 15mA lightbulbs
18-R03-10 1.5V 15mA red lightbulbs

Poly Scale paint
404100 Clear Gloss
414143 ATSF Silver
414146 ATSF Yellow
414212 Soo Line Red
414290 Flat Black

RadioShack
270-405 type N battery holder

Richmond Controls
EZ03 EZ-Lite for Kato F7

Sunrise Enterprises
N-499 speed recorder with cable

Tomar Industries
N-108 AT&SF *El Capitan* Conquistador tail sign

Top Flight
TOPQ0401 MonoKote Metallic Green

Chair car

Step-down chair car

9-1

CHAPTER NINE

Modeling Chicago's Dearborn Station

By Gary Hoover

Chicago's Dearborn Station was old and small, but it was home to some of the most interesting trains in the nation, including the Atchison, Topeka & Santa Fe's *Super Chief*. Here, locomotives and buildings are illuminated for night operation, an important feature on my HO layout.

Photos by the author

Imagine you're a traveler in the 1950s, making your first trip from sunny southern California to Chicago. You're on a luxury passenger train, the Atchison, Topeka & Santa Fe's *Super Chief*, and the trip has been first-class all the way. As your gleaming stainless-steel sleeper picks its way through the maze of turnouts and double-slip switches leading into Chicago's Dearborn Station, **9-1**, you notice switchmen manually aligning the route and occasional small shelters that protect the workers during Chicago's bitter winters.

As your train rounds the curve and heads into Dearborn's trainshed, you pass a few nondescript brick warehouses and several battered covered platforms full of sacks, mail pouches, boxes, and just about everything else imaginable. You're thrilled to spot some Wabash and Erie E units. Then, a Chicago & Western Indiana RS-1 flashes by on the next track, doing some switching duties.

Suddenly, the brightness of day turns to shadows as the *Super Chief* enters a dimly lit trainshed and slows to a stop.

You disembark onto a well-worn wooden platform. Entering the main station, you find that a short walk across the waiting room takes you through the main doors to Polk Street where a row of taxis are idling at the curb. You're surprised that such a luxurious trip has ended in such modest surroundings.

The charm of Dearborn Station

Dearborn's dour Gothic architecture, hemmed in by gritty freight houses and loading docks, was hardly what one would expect to find when arriving on the *Super Chief* in the railroad capital of North America, **9-2**. Although the station lacked the elegance and comforts of neighboring Grand Central and LaSalle Street stations, Dearborn had a charm all its own. It's the vivid contrast between spotlessly clean passenger trains and their dingy port of call that makes modeling Dearborn Station in the 1950s so interesting, so challenging, and, above all, so rewarding.

Modeling Dearborn Station and its surrounding buildings is a mid-level project – it's not exactly easy, but it's not very hard either. You'll need some time and a bit of research as you won't find ready-to-build kits for the structures. Even your best kitbashed and scratchbuilt buildings will probably be only good approximations due to the size of the prototype. Still, with a bit of patience and a dash of motivation, building an eye-catching Dearborn Station scene – or any complex city scene – is certainly possible.

Before tackling the actual construction of Dearborn Station, let's take a brief look at its history and note some of the distinct features of the 1950s that gave Dearborn its beloved, if slightly shabby, charm.

Chicago's oldest and smallest

In 1883, Cyrus Eidlitz completed plans for Dearborn Station, a combination of German Romanesque and Gothic styles. Two years and a half-million dollars later, Dearborn Station opened its doors to passengers. From the mid-1880s until the last intercity train departed on April 30, 1971, the station served millions of travelers.

Dearborn was the smallest of the six major railroad stations that would be built in Chicago. Red brick formed the exterior of the station with moldings and trim made from red terra cotta. The distinctive covered trainshed, constructed from wrought-iron columns and trusses and covered by a corrugated-iron roof, began at the back of the main station complex and ran 600 feet south with 10 tracks under cover. The Fred Harvey Co. lunch counter, dining room, and cocktail lounge were surely the crown jewels of the station. Here, travelers could relax while waiting for their trains and enjoy Harvey's world-famous cuisine.

Elsewhere in the station, travelers found the usual amenities such as ticket counters, baggage handlers, offices, and a newsstand, **9-3**. After World War II, a second-floor waiting room was added where large picture windows provided a view of the arriving and departing trains.

Landlords and tenants

Santa Fe's elegant trains are so closely associated with Dearborn that one

A warbonnet Santa Fe E6 and a handsome Wabash E7 provide a stark contrast to the otherwise gritty surroundings of Dearborn Station. The covered platform was scratchbuilt from Evergreen and Plastruct shapes using plans available from the Santa Fe Railway Historical & Modeling Society.

9-2

could be excused for thinking the Santa Fe owned the station. Actually, it was the relatively obscure Chicago & Western Indiana RR that ran the show. The C&WI was jointly owned by the Chicago & Eastern Illinois, Grand Trunk Western, Wabash, Monon, and Erie. The Santa Fe didn't arrive on the scene until late 1887, and then only as a tenant with a 999-year lease. In 1890, seven roads served the station with 122 trains and 17,000 people passing through each day.

In the 1950s, the Santa Fe may have been Dearborn's most recognizable railroad with its sparkling clean *Super Chief* led by the handsome red-and-silver warbonnet F units, **9-4**. However, the other roads are equally interesting.

For example, the C&WI performed switching for the owner roads. In the 1950s, the task was assigned to several C&WI RS-1s painted in the railroad's attractive dark green accented by bright yellow striping.

The Erie provided connections to the East Coast. Its trains took longer to get there than rivals operated by New York Central and Pennsylvania RR, but Erie's tickets cost less. In 1951, the Erie introduced attractive two-tone green E8s followed shortly by matching passenger consists for the *Erie Limited*. In early 1957, the Canadian National-owned Grand Trunk Western had the distinction of being the last Dearborn railroad to operate steam power on regular service.

The Monon served Indianapolis and Louisville with its stylish crimson, gray, and white F3As, while the Wabash featured dome car service to St. Louis pulled by its elegant blue, gray, and white E units. And finally, there were the trains of Dearborn's tenant, the Santa Fe. From its top-of-the-line *Super Chief* to the somewhat ragtag *Grand Canyon*, the Santa Fe streamliners provided Dearborn with a "Santa Fe All the Way" connection to the West and Southwest.

Unlike Dearborn's owner roads, which used the C&WI to perform switching, the Santa Fe opted to do its own. Distinctive Alco RS-1s in the classic black-and-silver zebra-stripe scheme did the honors, **9-5**.

By the mid-1950s, a Santa Fe steam locomotive in Dearborn Station was a rare sight. In summer 1953, many of the 45,401 Boy Scouts attending the National Jamboree in Irving Ranch, Calif., departed Dearborn on several special trains powered by Santa Fe steam locomotives. Then, on August 1, 1953, Santa Fe Pacific 3435 pulled out of Dearborn with train no. 9, the *Kansas City Chief*, the last Santa Fe to steam out of Dearborn Station.

Beauty in the eye of the beholder

While the Dearborn Station complex was probably the focal point for the general public, the decrepit structures surrounding the station are even more interesting and richer in character from a modeler's viewpoint.

Inbound and outbound baggage and mail buildings stretched from the east side of the main station and south along Plymouth Court and State Street. Flanking the station tracks to both the east and west were several tattered and worn freight platforms under sagging roofs.

On their way to board their trains, passengers could grab some reading material at this colorful newsstand operated by the Fred Harvey Co.

9-3

A pristine Santa Fe F3 is impressive to men and boys alike. A boy and his father catch a quick look at their warbonnet-painted locomotive before boarding the *Super Chief* to the West Coast.

Modeling a '50s prototype

Fortunately, a fair number of black-and-white and color photos of Dearborn Station and its surrounding structures exist. (See page 57 for some suggested sources.) Historical societies and Internet discussion groups are also excellent sources of information.

In any prototype-based modeling project, the ideal situation is to find photos for the exact time you wish to model. In reality, that rarely happens. I'm modeling August 29, 1951, for my layout, so I carefully compared photos taken before and after August 1951. If the same structure appears in a 1948 photo and again in a 1952 photo, it was certainly present in 1951!

Often, a good set of before-and-after photos is just not available. That was the case with the large Lee sign to the left of my Dearborn Station scene, **9-5**. An October 20, 1950, photo shows the building without the sign. The earliest photo I could locate that shows the sign was taken in September 1953. So, the eye-catching sign was painted sometime between late 1950 and mid-1953.

Who turned out the lights?

Night operation is an integral part of the operating philosophy for my railroad. The railroad's atmosphere magically changes when the incandescent layout lighting is dimmed and the interior building lights and signals can be clearly seen. To create the impression of a lively nighttime scene, most of the buildings on my layout needed to be at least partially illuminated.

Take a close look at any building at night, and you'll notice some rooms are lit and others are not. Applying that same observation to the buildings on your layout adds realism.

Preparing a building for night illumination is simple and requires only a few extra steps. When the building is basically complete but before it's painted and weathered, I make a test for light leaks by placing a 12-volt bulb inside the building and viewing it in subdued light.

In some cases, the plastic walls prevent the interior light from "glowing" through. In other cases, such as with structures built with thin wood walls, the interior will need a good thick coat of black paint. Small light leaks can be sealed with a dab of plumber's putty. When I have all of the light leaks patched, I paint and weather the structure and install the window glazing.

The final step is to decide which rooms to illuminate. For background buildings, I've found that the easiest and quickest way to make a window dark is to simply tape a piece of black construction paper to the backside of the window glazing. On buildings close to the viewer, black paper placed directly against the glazing might be too noticeable, so I make a small "room" using cardboard. A few interior details may be all that's necessary to give the illusion of a full interior.

For buildings or structures that may occasionally need to be moved for track cleaning or photography purposes, I connect the building wiring to the main lighting bus using a small pin plug.

9-5

Whether the sign is correct for August 1951 remains a mystery to me.

Vehicles, clothing, and types of passengers are good ways to reinforce a scene's time frame. Look at photographs of passengers traveling by rail in the 1950s, and you'll find that nearly everyone was dressed very nicely. Men were usually in suits, and women wore suits or dresses. Hats were common for both. In the early '50s, the Korean conflict was raging, and many uniformed servicemen and women were traveling by train.

More subtle items that establish the era include rolling stock, locomotive types, paint schemes, and sometimes even the roof geometry of buildings. Observers who have not studied the specific prototype probably won't notice these details, but being as correct as possible is what prototype modeling is all about.

Planning the station scene

As when building almost any model railroad layout, I had to use selective compression. In HO scale, Dearborn Station and the surrounding area could fill my basement, and I still wouldn't have enough room to model it to true scale. Fortunately, a very good representation of the prototype is possible by selectively compressing the scene. I chose to model the most prominent buildings and then scaled them back, **9-6**. I figure I've done my job well if an unprompted visitor exclaims at first glance, "That's Dearborn!"

I allocated a 4' × 14' area for my station. I needed to include some build-

With a building-sized Lee ad in the background, the *Super Chief* discharges its passengers, while Chicago & Western Indiana and Santa Fe RS-1s perform switching chores. The C&WI switched for all of Dearborn's owner railroads.

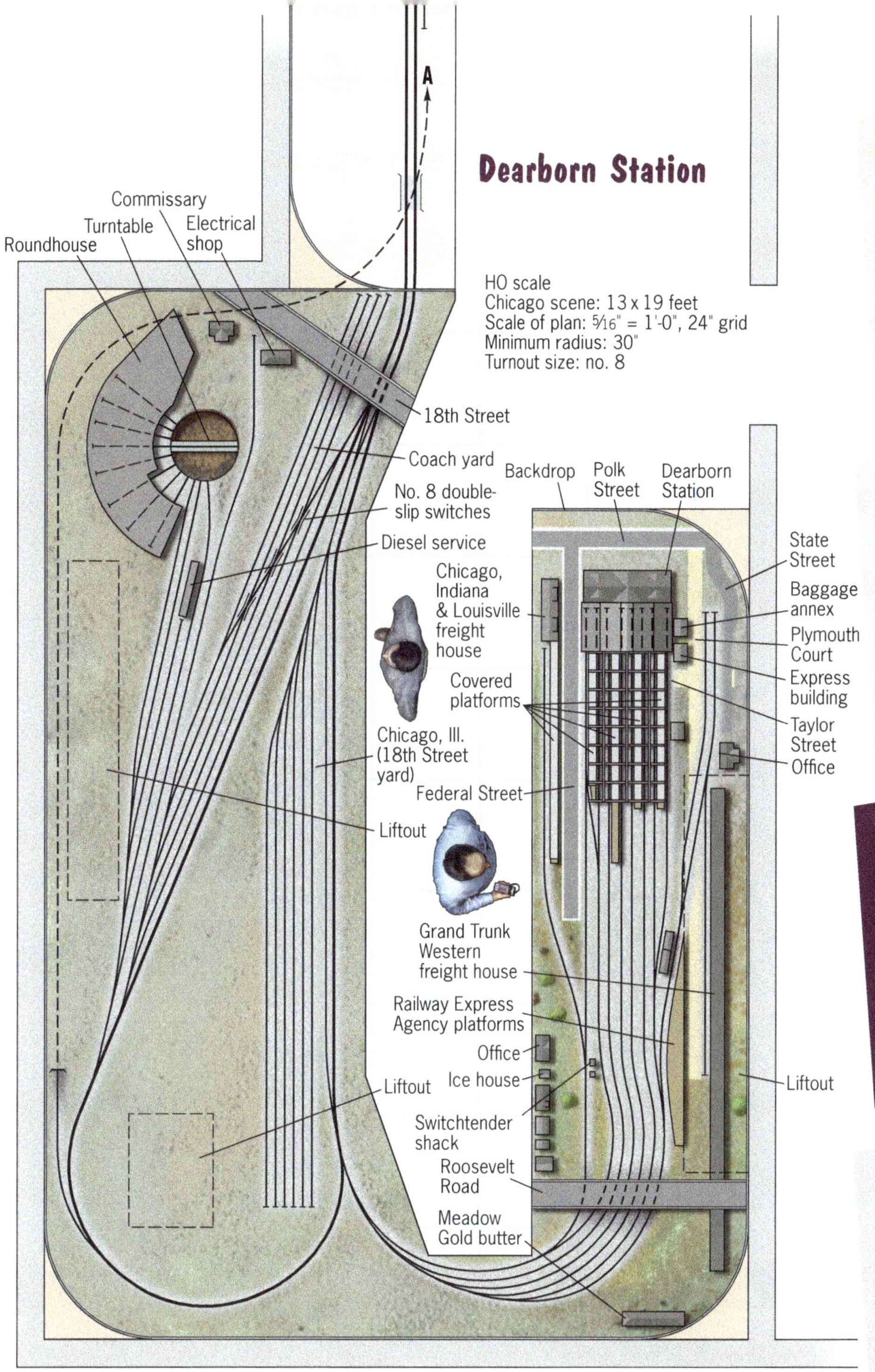

ing flats against the backdrop on the north side of Polk Street, the station, the trainshed, the Roosevelt Road overpass, and some scenery south of Roosevelt Road.

I wanted to have the scene start at the aisle side with the Monon freight house along Federal Street and extend back to State Street with more building flats along the backdrop. The scene's depth meant that I had to include small removable hatches running lengthwise along the backdrop. With all of the structures, I had no problem hiding the seams.

Since the station and trainshed were the largest structures, I built them first. I purchased a copy of the original blueprints for the trainshed and platforms from the Santa Fe Railway Historical & Modeling Society (www.atsfrr.com). These plans, in addition to giving me the footprint size of the station, gave me a basis to work from in compressing the structures.

I started by making cardboard mock-ups of the station, trainshed, and platforms to arrive at my optimum size for each of the structures. With the final sizing of the station complex complete, the most challenging and rewarding part of the scene construction began – building the structures.

9-6

Scratchbuilding Dearborn

Nothing exists straight out of the box for modeling a Dearborn scene. It was obvious early in my planning that all the structures would need to be either partially or totally scratchbuilt.

Scratchbuilding a structure most always provides a better representation of the prototype, assuming you have enough data to work from. On the other hand, scratchbuilding can take a fair amount of time. In the case of the Dearborn scene, everything was going to have to be compressed anyway, and all I had to do was decide which buildings I'd model.

By comparing prototype photos with structures in the Walthers catalog, I found kits that had relatively close matches for window geometry and wall arrangement. In the end, I scratchbuilt nearly all the buildings along the Dearborn Station backdrop. I also scratchbuilt the station complex, trainshed, and passenger and freight platforms using styrene.

"Good enough" makes it possible

In his clinics, noted modeler Allen McClelland often uses the term "good enough." Those two words, and an understanding of the concept behind them, can make the difference between dreaming and planning a layout and actually completing a layout.

"Good enough" means spending most of your time on what is seen, while minimizing the amount of time spent on what is not seen. For buildings against the backdrop, I didn't need to install a rear wall or spend much time on elaborate painting and detailing of the structures.

On the flip side, I paid extra attention to the details of buildings near the front. That's why all of the foreground buildings in my Dearborn scene have complete interiors, including the sta-

With only a 4' x 14' space to model my downtown Chicago scene, space was limited, so I modeled only the structures most easily identified with the area surrounding Dearborn Station. In this view, you can see the covered platforms. The entire portion of the scene along the backdrop can be removed for easy access.

tion lobby. But even here, the good-enough principle applies.

Simply adding a floor, some interior walls, a few pieces of furniture, paper drapes, and a strategically placed person or two inside a building gives the illusion of a completely detailed interior. Best of all, this approach meant that I was able to complete my Dearborn scene in about six months.

Track, turnouts, and routes

Passenger operations on my layout follow the same general pattern as the prototype. Arriving trains pull straight into the trainshed. After discharging the passengers, a switch engine pulls the train from the station and takes it to the coach yard and engine terminal. Trains readied for departure are backed into the station using switchers. The road locomotives are added before departure. Given the length of most passenger cars and the necessity to back trains, I used Code 83 no. 8 turnouts and a minimum track radius of 30". With properly weighted cars, quality wheelsets, and careful attention to diaphragm clearances, operation is derailment free.

The digital difference

Digital Command Control (DCC) systems offer incredibly realistic locomotive control and sound features. But using stationary decoders to control turnouts is one of the lesser-known benefits of DCC.

Why install a stationary decoder when a simple electrical toggle switch would do the same thing? The answer is "routes." A route macro is a way of specifying a string of turnouts that need to be properly aligned to get a train into a specific destination track. Many DCC systems and even some stationary decoders have routes as one of their standard features.

With a few programming steps, you can specify all of the turnouts that need to be open or closed for a specific route. After that, simply entering the particular route identification number sets the proper turnout position.

Placement, backdrops, and details

As I completed each building, I temporarily placed it in position for a final size check. When I had all of my structures assembled and painted, I applied weathering using pastels.

I added my backdrop details next. I used pieces of commercially available scenes for the majority of my backdrop. In areas where a specific prototype building was needed, I enlarged a reference photo from a book or magazine and applied it to the wall using spray adhesive.

The majority of my ground cover consists of real dirt applied with a fine sifter and secured by a light mist of diluted white glue applied with a spray bottle. I airbrushed all of the track with thinned Floquil Railroad Tie Brown. Next, I brush-painted the sides of the rail with a mix of 90 percent Floquil Rail Brown and 10 percent Engine Black. After ballasting the track, I used a brush to apply thinned Polly Scale Oily Black in select spots along the center of the track.

Day or night, the Dearborn Station scene is a major focal point of my Santa Fe layout, and its gritty charm still appeals to visitors of all ages.

References and suggested sources

Better data means better results. Here are some sources I used:

Bob Kosic, "Dearborn Station," *The Warbonnet* magazine, Second Quarter 2000, Santa Fe Railway Historical & Modeling Society, Midwest City, Okla., www.atsfrr.com

Lloyd Stagner, *Santa Fe in Color*, 1992, Morning Sun Books, Edison, N.J.

David R. Sweetland, *Wabash in Color*, 1991, Morning Sun Books, Edison, N.J.

Bill Marvel, *Santa Fe – all the way, Volume 1*, 1998, Morning Sun Books, Edison, N.J.

David R. Sweetland, *Erie Railroad in Color*, 1991, Morning Sun Books, Edison, N.J.

Joseph Welsh, *Passenger Trains of Yesteryear, Chicago Eastbound* and *Passenger Trains of Yesteryear, Chicago Westbound*, 2002, Kalmbach Publishing, Waukesha, Wis.

Bob's Photo, Ansonia, Conn., 203-734-6666

CHAPTER TEN

Freight equipment and operations

By Robert S. McGonigal

Team tracks were flanked by paved roadways to allow vehicles to reach freight cars. Beyond the team tracks in this Chicago scene is a small yard for sorting and holding cars. C&NW photo

Freight operations of the 1950s offer much variety for the model railroader. Rolling stock was shorter than it is today, so modelers can fit more cars into compact areas. Similarly, the greater prevalence of smaller shippers permits one to have more realistically sized industries on a layout. Trains were shorter and tended to be more varied in their makeup.

Many structures that are absent today, such as cattle pens, icing stations, and freight houses, were in use. More than 100 major railroads were in operation, so cars of many liveries could be seen – and they looked better too, as graffiti and corporate indifference had yet to take their toll.

Passenger trains were the most glamorous and publicly accessible aspect of 1950s railroading, but on all but a few carriers, they were dwarfed in importance by what has always been the industry's primary business: freight. Early in the decade, line-haul railroads produced around 616 billion ton-miles (one ton carried one mile, a standard measure of traffic) each year. By contrast, passenger-miles (one passenger carried one mile) averaged just 33 billion per year. At the same time, the 2 million freight cars on the nation's rails outnumbered passenger rolling stock nearly 50 to 1. By 1960, the gaps in both categories had widened.

Thanks to economic growth, total freight volume is much higher now than it was in the 1950s, but railroads of that era had a bigger – albeit shrinking – share of the pie. At the start of the decade, 56 percent of all intercity freight in the United States moved by rail; this figure slipped to 44 percent in 1960, and today stands at 42 percent. (Rail is gaining ground again, having bottomed out at 37.5 percent in 1980.)

Typical work days

Freight railroading in the '50s was conducted on a more decentralized basis than what we know today. Cars were smaller, trains were shorter, yards were more numerous, and there was a greater emphasis on local business.

For more than a century, rail was the primary means of moving products. Most major shippers and receivers had a siding or two, with loading/unloading facilities tailored to the types of cars it required. Big industries had entire yards and internal rail systems of their own. For customers without direct rail service, team tracks were available for public use, **10-1**. Pavement next to these sidings allowed access by wagon or truck. Freight houses and less-than-carload-lot (LCL) services also catered to off-line customers.

All those individual customers meant that thousands of times each day local trains, often called way freights, set out from yards to exchange cars with on-line businesses. Such a train might set out hopper cars of coal for a firm's power plant and pick up the empties, then position, or "spot," a couple of cars of raw materials at the loading dock and pull out cars bearing the industry's product. A bit farther down the line, it might deliver tank cars to a local fuel distributor and then flatcars to a lumberyard. Local freights also carried "peddler" or way cars that dropped off and picked up LCL shipments along the train's route, **10-2**.

At yards, cars from local freights would be classified by destination. Some might be grouped for inclusion in another way freight, but longer hauls were more common. Most cars would be switched into trains headed in the direction of their final destination hundreds or thousands of miles distant. On a cross-country journey, a car might be classified several times, as well as interchanged from one railroad to another.

Yards were of two basic types: flat or hump. Flat yards could contain a handful of tracks – or dozens. Cars were switched by a locomotive pulling a string, or "cut," of cars from one track and then pushing single or groups of cars into other tracks. The switch engine might have to reverse direction many times until the entire cut was classified. Switches were generally hand-thrown.

Less numerous but bigger than flat yards, hump yards contained dozens or scores of tracks. The key was a small hill, the hump, over which an engine would slowly push strings of cars. At the apex of the hump, cars not bound for the same destination would be uncoupled by crewmen walking beside the train. The cars would then roll down the hump to the bowl of the yard,

10-2

A Wabash local freight has stopped at a grade crossing, so a boxcar (straying from the auto service for which it's lettered) can have its load transferred to a grain elevator's truck. Linn H. Westcott photo

10-3

While general freights incorporating a variety of car types might be more interesting to look at, solid trains of hoppers (like this Virginian Ry. coal drag), tank cars, reefers, or stockcars were common. D. Wallace Johnson photo

10-4

At Addison, N.Y., F units lead a through freight west as a B&O 2-8-0 prepares to depart with a branchline freight. A variety of 40-foot boxcars is shown. The Lackawanna RR boxcar (at far left) is lettered to promote the road's *Phoebe Snow* passenger train. David R. Connor photo

routed to the proper tracks by remote-control switches. In early hump yards, car speed was controlled by brakemen riding the cars; by the 1950s, most hump yards had mechanical retarders that pressed against the car wheels to slow them down.

In many areas, a local commodity dominated all others. In farm country, way freights gathered boxcars full of grain from elevators located six or eight miles apart (covered hoppers weren't adopted for grain hauling until the 1960s). California, source of many perishable crops, had a nearly insatiable appetite for refrigerator cars to whisk produce to eastern markets.

In the iron ranges of Minnesota and Upper Michigan, legions of stubby ore cars moved iron ore, the lifeblood of the steel industry, to Great Lakes docks. And of course there was coal, which accounted for about half of all rail freight tonnage; long strings of hoppers full of black diamonds were part of the landscape in West Virginia, Pennsylvania, Kentucky, Illinois, and other states, **10-3**.

Car types

The most common freight car types of the 1950s were boxcars and hoppers. In 1956, the composition of the national car fleet looked like this: boxcars, 37 percent; hoppers, 27 percent; gondolas, 14 percent; tank cars, 8 percent; refrigerator cars, 6 percent; flatcars, 3 percent; covered hoppers, 3 percent; and stockcars, 2 percent.

Freight cars are usually described by the weight of cargo they can carry or by their inside length. Most cars of the 1950s had a capacity of 50 to 70 tons, although toward the end of the decade iron-ore hoppers capable of hauling 95 tons began to appear. (Today, loads can tip the scales at 140 tons or more.) At the same time, about one-quarter of the car fleet was 30 or more years old, dating from an era when average capacity was less than 45 tons.

The workhorses of the era were boxcars with an inside length of 40' 6" or 50' 6", **10-4**. Hopper cars came in two- and three-bay designs that measured from about 31 feet to 41 feet. Gondolas ranged from about 41 feet to 66 feet. Most cars were of all-steel construction, although nearly 185,000 composite (steel framing, wood sides) box, hopper, and gondola cars were still around in 1954. Nearly all of the 39,000 stockcars in service that year had wood slats.

More than a dozen suppliers, and several railroads, built freight cars in the 1950s. Among the commercial firms were American Car & Foundry, Bethlehem Steel, National Steel Car, General American Transportation, Thrall Car, and the biggest: Pullman-Standard. In 1957, Pullman-Standard's line included PS-1 40- and 50-foot boxcars, PS-2 30- and 40-foot covered hoppers, PS-3 50- and 70-ton hoppers, PS-4 54- and 56-foot flatcars, and PS-5 52- and 65-foot gondolas.

The 1950s saw a number of innovations in freight-car types and technology. The use of roller bearings on freight equipment, universal today, gained momentum, although by 1957, just 1 percent of the car fleet had been so equipped. Mechanical refrigerator cars began to replace ice-cooled cars in the late 1940s, but in '57, they still made up only 1 percent of all reefers.

The carriage of truck trailers and containers on flatcars (intermodal, which is a major part of today's rail-freight business) was practiced by just six railroads in 1952; most used standard cars that they had modified with side rails and tie-downs. Intermodal got a big boost with the 1956 formation of Trailer Train, a company that owned the specialized flatcars required for piggyback service and provided them to the carriers.

Trailer Train's first order was for 500 75-foot flats capable of carrying two of the then-standard 35-foot trailers. Several other intermodal systems were tried in the '50s, including Chesapeake & Ohio's Railvan (a 1959 forerunner of today's RoadRailer) and New York Central's 1958 Flexivan, which allowed the transfer of containers to flatcars without the use of cranes.

10-5

In the '50s, most freight crews consisted of an engineer, fireman, head brakeman, rear brakeman, and conductor. Here, three crewmen are with a wooden New York Central caboose. Jim Shaughnessy photo

Rolling billboards

The 1950s were something of a golden age for freight-car liveries and lettering. Many roads replaced durable but drab boxcar red on at least some of their cars with eye-catching, multicolor liveries: black and silver on the Central of Georgia, blue and black on the Boston & Maine, and red and gray on the New York Central. Freight cars may have gotten dirty, but they were free from the large-scale graffiti that disfigures so much of today's rolling stock.

The external posts characteristic of boxcars today had yet to be adopted, meaning that many car sides were as flat as billboards – which is exactly how some roads treated them. Soo Line, Great Northern, and Western Pacific were among the carriers that proudly spread their names across their cars at unprecedented size. Other roads complemented their names with slogans such as "The Old Reliable" (Louisville & Nashville), "Main Line of Mid-America" (Illinois Central), and "Everywhere West" (Burlington Route). Fondly remembered are the boxcars, reefers, and other cars that advertised their owners' passenger trains.

Bringing up the rear

Perhaps the most appealing aspect of 1950s freight trains – at least for railfans and modelers – is that they all had cabooses. Even better, there were nearly as many caboose designs as there were railroads, and the cars' colorful nicknames – hack, crummy, waycar, buggy, van, cabin – seemed nearly as numerous. Most carriers liked rooftop cupolas for their cabooses, but many specified bay windows. Cupolas could be centered, offset, high, low, or even streamlined, and bay windows also came in a variety of styles.

Some crummies had neither cupola nor bay windows. Transfer cabooses were used for slow-speed, short-haul movements between yards of different railroads – runs with less need for observation of the train from the rear. These cars often had large end platforms to give crewmen a better place to ride during the frequent backup moves made by transfers, **10-5**.

Many wood-bodied hacks were still in service in the 1950s, and there were even a few four-wheel "bobbers" left. Cabooses weren't always at the rear end. A way freight's frequent switching might result in the caboose being in the middle of the train, right behind the engine, or even in front of the engine for part of the run.

The typical view of early 1950s freight trains tends to be that of monochromatic boxcar red consists. By the end of the decade, billboard paint schemes, attention-getting color combinations, and a wide variety of car types, punctuated by the (usually) bright red crew car that every child of the day watched for, made the 1950s a modeler's delight.

CHAPTER ELEVEN

Replacing war-weary freight cars

By Mont Switzer

11-1

When you model the steam-diesel transition era, your fleet will include an interesting mix of steel and composite cars such as these models.

Bill Zuback photo

If you were trackside in the 1940s, you would have seen freight cars in Tuscan red, black, and gray and an occasional reefer in orange or yellow. But drab colors weren't the only distinctive characteristic. Prior social and economic conditions contributed to the appearance of America's trains at mid-century. A railroad-car building boom in the 1920s put large numbers of steel-underframe cars and some all-steel cars in service. But few new cars were built during the Great Depression, leaving the railroads ill prepared to handle the record tonnage moved during World War II. Wartime rationing of steel severely limited new freight car production, and the cars that were built then often followed earlier composite construction designs.

Once the material restrictions were lifted, new steel-car building increased as war-weary cars were replaced with new, innovative designs, and postwar freight trains became colorful showpieces of the railroads' ability to rebound.

Perhaps the best part of modeling the steam-diesel transition era is that most of these changes took place over a number of years. You can be comfortable about building and running a wide variety of equipment representing the end of one era and the beginning of another with the promise of greater things to come, **11-1**.

Boxcars

Composite wood-and-steel boxcars were rebuilt in steel or replaced with new, larger-capacity steel cars. A 40-foot car with 6-foot door openings was the most common size built. Inside height was increased 12 to 18 inches to a maximum of 10' 6". Some cars from the 1930s were modified to increase cubic capacity and inside ceiling height, and they remained in use through the '40s, **11-2**.

Single- and double-door 50-foot boxcars were also being built as customers demanded more cargo space for lighter commodities. Welded construction and larger door openings were the rule by the end of the decade.

11-2

Composite cars like this Atchison, Topeka & Santa Fe car had its roof raised to increase capacity. Photos by the author unless noted

11-3

Steel boxcars from the '30s were updated with ad slogans that promoted passenger or freight services.

11-4

Refrigerator cars were often rebuilt with steel ends and roofs but retained their wood sides, which helped insulate the cars.

11-5

All-steel construction made reefers more durable. The ends and roofs of the cars were usually black or brown.

11-6

Steel, 50-ton twin hoppers all but replaced the gondola in the coal-hauling and bulk-materials trade.

11-7

Composite hoppers were scrapped or rebuilt in steel. Two-bay hoppers were commonly used for shipping coal to small dealers.

Boxcars painted during the 1950s became billboards for their owners as railroads applied slogans promoting their passenger trains and fast freight services, **11-3**. Through the 1940s and '50s, as low-cost, brightly colored pigments came along, the drab, soot-stained freight trains of the steam era became quite colorful. Before long, however, fewer railroads advertised passenger trains on these colorful cars as that business declined.

Refrigerator cars

Because of wood's excellent insulating characteristics, the move to steel construction in refrigerator cars occurred more slowly than it did in boxcars, **11-4**. Older reefers were rebuilt with steel ends and roofs, while new refrigerator cars featured all-steel construction, **11-5**.

Mechanical refrigerator cars became increasingly popular due to their ability to provide more evenly controlled temperatures, including those required for a 1950s innovation in food marketing – frozen food.

The insulated boxcar – essentially a bunkerless refrigerator car – was a close cousin to the ice reefer. It became popular as more and more customers shipped canned beverages and products that required protection from temperature extremes but not temperature control.

The traditional vivid yellows and oranges remained on refrigerator cars, but ends and roofs were usually brown or black. Some companies experimented with roofs painted in reflective colors.

Refrigerator cars were typically owned by companies that specialized in handling perishable shipments. About 21,000 cars were privately owned, while nearly 97,000 cars were in the fleet of the larger reefer-operating companies owned by railroads.

Hopper cars

In the 1950s, hopper cars were built by the hundreds of thousands. The majority of these cars were self-clearing, two-bay, all-steel hoppers of 50-ton capacity, **11-6**. Older two-bay cars of prewar composite construction were rebuilt with new steel sides or scrapped, **11-7**. Two-bay cars were ideal for single-car shipments to small coal dealers.

As fuel oil, propane, and electricity replaced coal, larger three-bay

11-8

Easy to load and unload, new covered hoppers began handling dry bulk materials.

11-9

Worn-out composite gondolas were on their way out, replaced with all-steel construction.

11-10

New steel gondolas were more rugged in hard service, but as a result, were subjected to damage from various loads and rust.

11-11

Tankers like this 8,000-gallon car, a size built since the 1920s, no longer met the needs of changing industries.

11-12

Increased length (here, 53" 6'), stronger cast-steel frames, and larger bearings helped increase the load capacity of flatcars.

hoppers with outside posts and welded construction designed for bigger loads began to appear.

Oxide red and black were the colors of choice for most hopper cars. Lettering was typically spartan with reporting marks and heralds in white.

Most railroads had at least a few hoppers, but the largest fleets were owned by the major eastern and southern railroads that originated most of the coal traffic.

Covered hoppers

Covered hoppers became popular for hauling dry bulk material, **11-8**. Manual labor was reduced as new material-handling systems quickly loaded and unloaded these cars. Most early covered hopper cars were two-bay, 50-ton, all-welded designs, but larger cars were soon being built.

The most common color scheme was light gray with black lettering. However, some railroads stuck with their standard oxide red or black with white lettering.

These cars were specialized enough that large fleets were privately owned by companies that, in turn, leased them to shippers and/or railroads. As customers embraced this new technology during the 1950s, some railroads also purchased covered hoppers in large quantities.

Gondolas

As composite boxcars were being replaced, so were composite gondolas, **11-9**. New gons were made entirely of steel. Gondolas carried a multitude of loads – just about anything that could be lifted in and out of the car by an overhead crane, **11-10**. As with boxcars, the 40-foot size was most common, but longer all-steel gondolas began rolling over the lines with increasing frequency.

Paint and lettering on gondolas was similar to that on hopper cars – oxide red or black bodies and white lettering.

Although nearly all railroads owned several gondolas, most were owned by railroads that used them to serve customers having heavy industrial demands.

Tank cars

The large fleets of tank cars that once hauled petroleum products nationwide shrunk as pipelines were developed and the trucking industry expanded. The growth of the chemical industry and the resulting more-specialized needs led to the retirement of many older, riveted cars, **11-11**. A number of newer car designs came into use with higher-capacity tanks that were

11-13

Older flatcars were often converted to haul the new piggyback trailers on scheduled, fast freight trains.

usually welded and fitted with insulation. It took some time to make the transition, so both types of tank cars were seen in freight trains during the late 1940s and early 1950s.

Flatcars

The common 40-foot flatcar, mostly railroad-owned, disappeared as customers demanded longer cars with stronger steel frames for carrying heavier loads. Although flatcars represented a relatively small portion of the nation's car fleet, their visible cargos made them interesting. Flatcars mirrored gons and hoppers in having drab paint and lettering. Because of limited space on the sides, they typically carried only reporting marks and dimensional data.

Piggyback flatcars

When piggybacking (carrying highway trailers on flatcars) was introduced, railroads were reluctant to invest in new flatcars for this unproven service. Many refitted older flatcars to haul highway trailers on fast freight trains, **11-13**. It quickly became evident, however, that one trailer per car was inefficient, and by 1955, flatcars were being built to transport two 35-foot trailers.

Many of the new piggyback cars were built for Trailer Train Co. (TTX), an equipment leasing company jointly owned by its member railroads. In 1956, TTX operated an extensive fleet of 75-foot, two-trailer flatcars; in 1957, a fleet of TTX 50-foot cars came along to handle the newly legalized 40-foot trailers. For a time, TTX flatcars were traveling nationwide in front of, and behind, the railroad-owned flatcars. By the early 1960s, however, piggyback trailer leasing was becoming more common.

In developing their own piggyback trailer fleets, the railroads had to learn their customers' needs, as well as learn how to construct trailers that could withstand the rigors of rail travel. Considerable thought also went into the paint schemes railroads applied on their new, highly visible highway trailers. Nearly any color imaginable was possible, but many railroads went with their passenger train colors on the trailers. Trailer ownership varied, with railroads owning some or all of their trailers, especially those with the more elaborate paint schemes.

Early trailer-pooling companies provided trailers to any railroad upon demand. Trailers leased for long terms were even marked with a railroad's heralds, usually placards riveted on to keep costs down and to allow easy changes when the lease terminated. Pool trailers on short-term leases carried their owner's reporting marks.

11-14

New all-steel cabooses included features and interiors designed to enhance crew safety and comfort. Jim Hediger photo

Cabooses

When compared to the new all-steel cars, wooden cabooses still in service at mid-century were quaint throwbacks, **11-14**. The typical dark red or brown was giving way to brighter colors. For safety reasons, many railroads painted the car ends bright yellow or orange, and some railroads created eye-catching schemes for cars used in special services such as merchandise and piggyback trains.

Union agreements during the 1950s allowed some railroads to drop the practice of assigning cabooses to individual crew turns. Fewer cars were needed, as off-duty crews began staying in motels instead of in a caboose at distant terminals. Branch lines were being abandoned, and multiple-unit diesels pulled longer (and fewer) trains. All these changes reduced the demand for cabooses. Older wood cars were scrapped or demoted, leaving the newer steel cabooses to bring up the rear of road freights until they were replaced by end-of-train devices in the 1980s.

12-1

CHAPTER TWELVE

Hauling Appalachian coal

By John Armstrong

Despite the acquisition of diesel locomotives, many railroads still had a significant roster of steam in the mid-1950s. As an example, this pair of C&O Mallets move a coal train near Holden, W.Va., on a spring day in 1955.
Robert Collins photo

Suppose both the Chesapeake & Ohio's main line across the eastern Continental Divide and an important secondary Pennsylvania RR route toward tide-water somehow depended upon a jointly owned short line to complete their routes. What a place it would be for the railfan to watch everything from mainline passenger trains to coal drags rolling through spectacular scenery behind an incredible variety of steam and first-generation diesel motive power! That's the premise of the O scale Blue Ridge Midland's revision of eastern trunk-line geography.

By 1955, most railroads that hadn't completely dieselized already were purchasing the new locomotives as fast as manufacturers could produce them. However, to cope with peak traffic, very large railroads, such as the Pennsylvania and the Chesapeake & Ohio, still had a significant number of steam locomotives on their rosters, **12-1**.

According to Lloyd Stagner's book *American Steam Finale 1954 to 1970* (South Platte Press), the C&O listed about 90 active steam locomotives in 1955, including twelve 2-6-6-6 Alleghenies operating in coal service, **12-2**. Steam was still prominent on various sections of the Pennsy in 1955 as well. Despite the fact that the PRR received 76 new diesels that year, it listed 206 steam locomotives in active service during the winter of 1955-56.

Although both railroads were largely out of the steam business by the end of 1957, you could convincingly model the Blue Ridge Midland as a steam-heavy, coal-hauling road in the early to mid-1950s.

Starting with staging

I designed the Blue Ridge Midland for a client, who had a 23' × 32' area to work in. The uncluttered space allowed the use of all four walls, which gave the design a lot of mainline action while maintaining 54"-minimum-radius curves.

There was already a platform along the back wall with a simple single-ended staging yard laid on it. The benchwork for this section was at a back-straining 34" height. It was low enough, however, to be covered with Midland City's yard, leaving enough vertical separation to provide adequate access to the lower level.

The staging yard becomes bidirectional when connected as the tail of a reverted loop – the Ohio loop on the plan. Trains leaving the main line at Cheat Junction (named for the West Virginia river) can be stashed away before or after being backed around the reverted loop. With a radius of 48", the Ohio loop can accommodate every type of rolling stock on the railroad except perhaps the most rigid-drivered steam locomotives.

Above the staging yard is Midland City, a subdivision point on both railroads with a long station platform, two freight yards, and a compact engine terminal with a turntable long enough to handle a C&O 2-6-6-6 Allegheny. Across from the station is a five-track yard with three tracks having double-ended access to all main lines. The shortest of the yard's stub-ended local tracks also doubles as a station track.

Coal gets its due

Descending from the southwest at a brake-testing 3.5 percent is a busy branch from the mines. It reaches Midland City through a connection at Coal Junction and has its own four-track yard for classifying loads to be picked up by mainline drags; two other tracks are available for assembling longer cuts of the 50-ton twin hoppers typical of this era.

C&O and most Pennsy trains go their separate ways on double-track main lines at Aberfoyle Junction. A crossing providing access from the freight yard is there, not only because it looks nice but also as a source of the *whumpety-whump* of passing O scale wheels.

Chesapeake & Ohio's 2-6-6-6 Alleghenies, such as this one seen at Ronceverte, W.Va., were the heaviest reciprocating steam locomotives ever built, making for some exciting high-iron action. Running big engines like these is a primary goal of the O scale Blue Ridge Midland RR. William P. Price photo

12-2

FREIGHT TRAINS

Blue Ridge Midland RR
Lower level

O scale
Layout size: 23'-2" x 32'-2"
Scale of plan: ¼" = 1'-0", 12" grid

Minimum radii:
Main line: 54"
Industry tracks: 36"

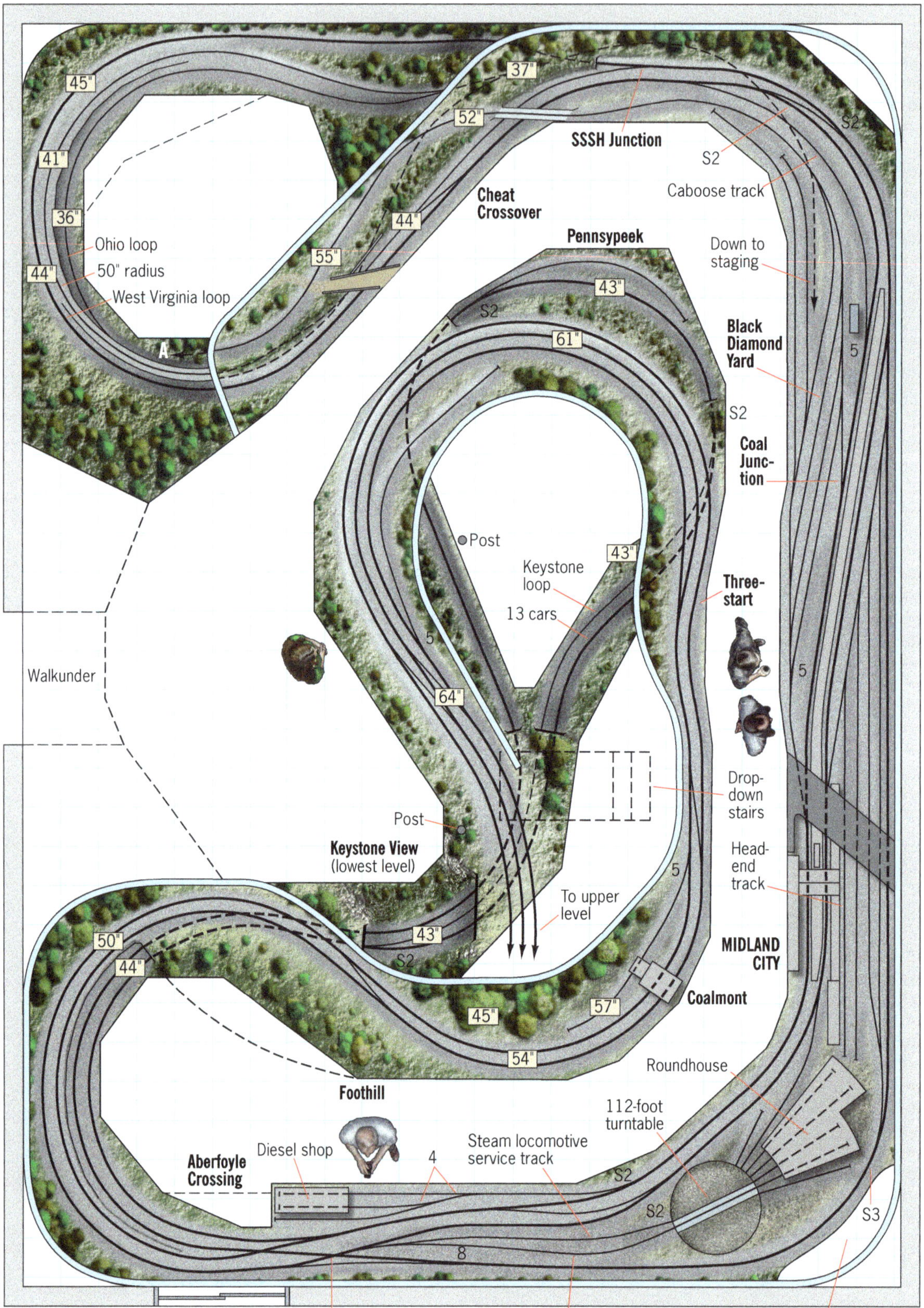

Upper level

Unmarked turnouts are standard No. 6
Other standard turnouts as marked:
S1 = 48"/58"-radius curved turnout
S2 = 48"/59"-radius curved turnout
S3 = 49"/59"-radius curved turnout
S4 = 63"/83"-radius curved turnout

illustrations by Rick Johnson and Theo Cobb

Schematic and profile

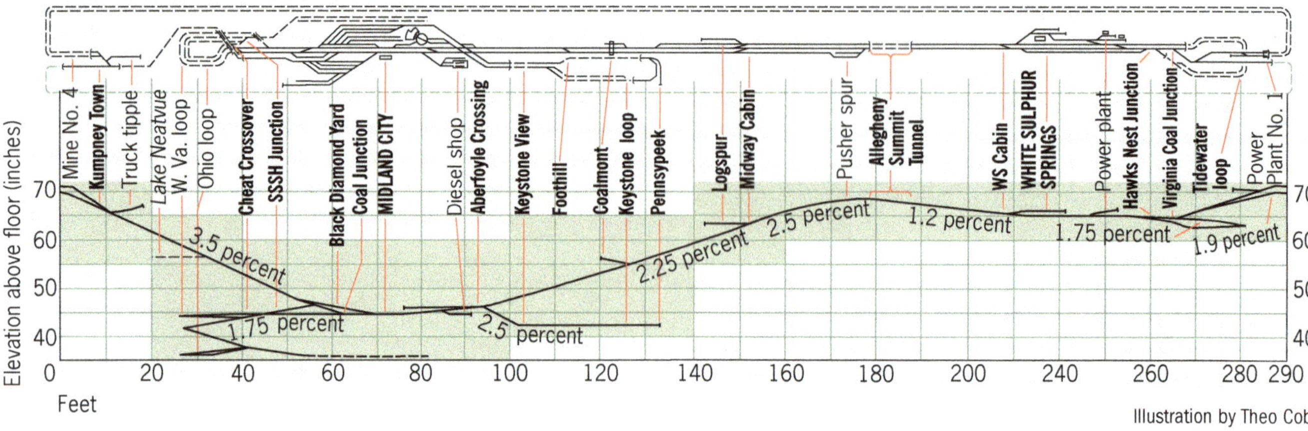

Illustration by Theo Cobb

Hiding those tracks

Plans like the Blue Ridge Midland compress a lot of tracks into tight spaces to give a good-size mainline run or create the illusion of having one or more branch lines. Often, however, the tracks for two different locations run near each other, so it takes some careful scenery work to hide and disguise them.

One technique, as shown in the illustration, is to make a scenic shell over the tracks to be hidden. In this example, foam insulation board is cut into support ribs, which are covered with several layers of plaster cloth. (A similar effect could be achieved by using screen wire or a web of cardboard strips to support the plaster cloth.)

When the plaster dries, you have a durable, yet hollow, shell that will support just about any type of scenery materials you wish to use. This way, a train can't be seen passing through any location on the layout more than once.

– David Popp, MR

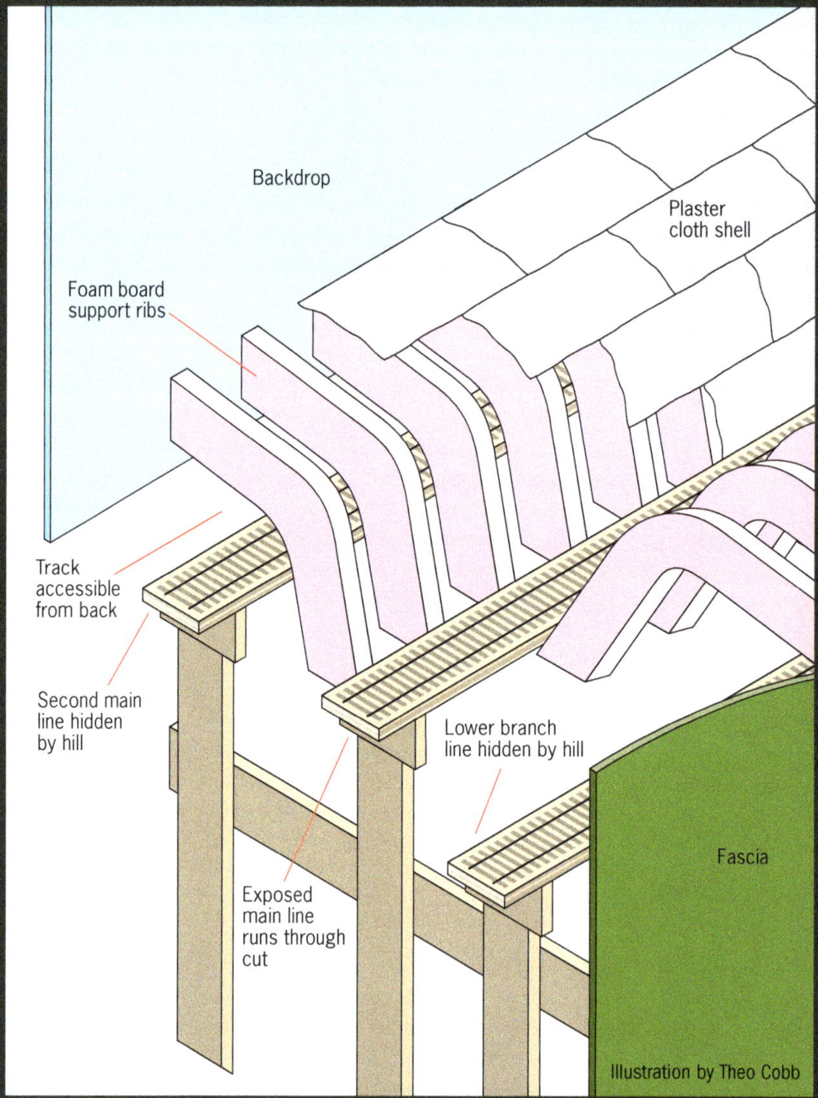

Illustration by Theo Cobb

12-3

An eastbound C&O passenger train holds to the south bank of the Kanawha River at Hawk's Nest, W.Va. The bridge in this photo is the inspiration for the Hawk's Nest Junction bridge on the track plan. Joe O'Hearn photo

The PRR line terminates in the Keystone loop. Once the upper C&O line is in place and its scenery completed, the loop is visible only in short vignettes at Pennsypeek and Keystone View. A layover track in the hidden section adds to the Pennsy's limited staging capacity, but the loop's biggest contribution comes early on, when it provides abundant loop-to-loop and out-and-back operating options while the C&O main line and other branchline extensions are being constructed.

Unmistakably Chessie

Leaving Midland City to the east, the mostly C&O main line starts its ascent to Allegheny Summit, attaining enough altitude to clear the walk-in entrance. The maximum grade is 2.5 percent, requiring heavy passenger-train power but usually no helper. The line gets to the upper level without a helix; from Threestart past Midway Cabin to the summit tunnel, there is an additional track for freight trains, which require helpers.

Once past the tunnel, the line descends gently into White Sulphur Springs. After passing a somewhat condensed facade of the classic Greenbrier resort hotel, the *Sportsman* and other C&O flyers enter a replica of the tracks and platforms of one of Chessie's most elegant small stations, including the short siding and spurs for setting out Pullman cars.

Beyond White Sulphur Springs, the main line terminates in a topside loop with layover siding, partially tucked under the horseshoe curve terrain. The exit from the loop is across the aisleway on an impressive through-truss bridge at Hawk's Nest Junction, reminiscent of a C&O situation in West Virginia, **12-3**. The bridge is at eye level, so it's only a slight duckunder.

Branches with a Mallet flavor

Perhaps the most distinctive feature of the plan is the alternate line from Virginia Coal Junction across the entryway and back down, via a loop around Lake Neatvue, to the west end of Midland City at Coal Junction. From Coal Junction until it quietly disappears behind a ridge in front of Mine No. 4, it's a heavy-duty branch serving mine after mine. Solid trains of empty hoppers head up the line all day, including the particular mine run that switches the truck tipple at Kumpenytown and Mine No. 4, and return with loads.

From Virginia Coal Junction until it disappears in the valley where Power Plant No. 1 is nestled, the alternate route represents an equally heavy-duty secondary main line. Thus, in addition to the passenger and merchandise traffic flowing through Midland City in both directions, there is also a realistic stream of eastbound black diamonds and westbound empties, making for a busy model railroad.

CHAPTER THIRTEEN

Moving produce to market

13-1

Pulling a string of loaded Pacific Fruit Express refrigerator cars, a Southern Pacific road switcher passes the orange trees that help keep the railroad busy.

Photos by the author

By Robert Smaus

If anything shouts California railroading in the 1950s, it's long strings of bright orange Pacific Fruit Express (PFE) refrigerator cars snaking though mountain passes. Headed east, they carried produce, or what railroaders call "perishables," as quickly and coolly as they could.

Within a few days, these loads of artichokes, citrus fruit, celery, lettuce, and other produce were being sold in eastern supermarkets. Even canned goods and a variety of frozen fruits and vegetables traveled cross-country in this manner.

Pacific Fruit Express had the largest fleet of ice refrigerator cars in the country, twice as many as its closest competitor, the Atchison, Topeka & Santa Fe. Other railroads and refrigerator car lines also had fleets numbering in the thousands, since the only way things got to market in the '50s was by train. Trucks hauled fresh produce only for short distances because the Interstate highway system wasn't built yet.

A crop of fun

These bright orange reefers were the reason I decided to model the Southern Pacific as it was in the early 1950s, **13-1**. (See chapter 2 for more on my HO layout.) The SP and Union Pacific jointly owned PFE, so I had visions of long reefer blocks being pulled by massive Cab-Forwards straining to get over the mountains. When I began to plan my layout, I quickly realized that the fun might actually lie in modeling the industries that used those strings of ice refrigerator cars.

To do that, I would have to plant the orange groves (see sidebar on page 78), pick the crop, pack the fruit, and ice the cars before they could be on their way. I could build an entire layout around this packing and shipping operation. I'd like to say that I made an in-depth study, but I simply started looking closely at photographs to get a feel for these agricultural industries and what elements I needed to model for a similar effect.

Much like an impressionist painter, I didn't need too much detail, just broad strokes that conveyed the flavor of the citrus industry. Only then did I do a little additional homework to find out how some of the structures were actu-

PFE's top perishable loads in 1953

Potatoes	85,000 carloads
Lettuce	66,819 carloads
Citrus fruit	34,622 carloads
Cantaloupes	26,836 carloads
Canned goods	25,014 carloads
Carrots	16,591 carloads
Tomatoes	15,979 carloads
Bananas	15,877 carloads
Celery	14,464 carloads
Grapes	14,159 carloads

ally built or how they worked. Later, I learned how the railroads actually handled the produce. That turned out to be almost as much fun as the modeling.

The most valuable crop

I decided to model the citrus business because that was the most valued and most noticeable crop shipped east. After picking, oranges were carefully placed in field boxes and hauled to the packing plant, first by horse and wagon and later by truck, **13-2**. At the plant, each piece of fruit was washed, waxed, and graded. In the early days, the fruit was wrapped in squares of tissue paper before being packed into crates.

Those who model other parts of the country can enjoy a similar hunt for an appropriate perishable commodity. If I were modeling the Northwest, I might have chosen apples, or if I lived in Maine, it would be potatoes. Every area of the country produces something. But in the '50s, citrus was king in Southern California, even though PFE actually carried more potatoes and lettuce than it did oranges and lemons. However, fields of lettuce or potatoes aren't as glamorous as the romantic images of orange groves.

Citrus fruit is a year-round crop, with Valencia oranges ripening in summer and navel oranges in winter. Lemons grow in both seasons near the coast, so it's always busy at a lemon packing plant.

13-2

Freshly picked oranges are trucked from the citrus groves to the packing plant in sturdy, shallow wooden boxes to prevent damage to the fruit.

13-3

This large, mission-style building is the Sunkist plant in Fillmore, Calif. It was the prototype for my model.

13-4

Although the model of the Sunkist plant was compressed, it still captures the prototype's mission-style construction.

FREIGHT TRAINS

13-5

A Southern Pacific switcher brings iced empties to swap for the outbound loads at the citrus fruit packing plant.

13-6

Box manufacturers operated in simple frame structures with lean-tos, which can be modeled quickly using sheets of basswood siding. It was common to find these factories next to the packing houses that used the crates they made.

Modeling a packing house

When I began modeling a citrus packing plant more than 10 years ago, I had numerous prototypes to choose from.

One, in Fillmore, Calif., intrigued me more than any other since it had an interesting mission-style covered loading platform, **13-3**. I modeled the structure from Strathmore art board, a heavy card material favored by early modelers.

At the time, I didn't have a layout to put it on, but it sure was fun to build the packing house, especially the little low-relief of a lemon on the front facade, which I made using a clay mold and some odd little plastic pellets that would melt in an oven to make the lemon. The big yellow fruit was a prominent feature on the actual building, and packing plants were often embellished this way since entire towns were built around this important industry.

My model is a big structure, with a footprint measuring 116 × 145 scale feet, so I can easily spot three reefers for loading with four on an adjoining track. The actual building is huge. I felt it was important to keep the model large, even if it was highly compressed, **13-4**. I wanted this plant to dominate the scene, so there's little doubt that my railroad has work to do.

The actual plant in Fillmore had four loading doors under the arches and a half dozen more in a later addition. Two additional parallel tracks were also used for loading. Plank bridges connected adjoining rows of cars, so workers could push hand trucks though one car to load another.

When produce is harvested, huge quantities have to be packed quickly and sent on their way. Because of a lack of space, I didn't try to replicate this high level of activity. I did, however, add a long storage siding next to my loading track, so quite a few reefers are in close proximity to the plant.

My packing house sits inside a 36"-radius curve, with the town off to one side. The forked spur to the plant branches off a siding in the town, which means I can switch the packing plant without fouling the main, **13-5**.

13-7

Icing took a lot of manual labor, first to deliver the 300-pound blocks and then to break them into chunks that would fit into the car's ice bunkers.

Building a box factory

There was enough room on the siding to add a related industry, so I modeled a box-making company. (See the April 2000 issue of *Model Railroader* for an article on building this plant.) These plants received lumber called "shooks" and shipped out some of the resulting wood packing boxes and crates, although most simply went next door, **13-6**. It was very common for Sunkist and other cooperatives to have their own box factories, although I modeled mine after an independent company in another town.

Box factories weren't as glamorous as packing houses, so inspirational photos are uncommon. Most of these structures are simple, rectangular buildings, with a lean-to occasionally added. I modeled mine quickly and simply with sheets of basswood siding and included an elevated foundation and dry-transfer sign to give it some character.

Being on the same siding as the packing plant, the box company adds interest to switching operations. To move empty reefers from storage to the loading area, cars of shooks must be moved out of the way. They can be pulled back into the town siding and left while the reefers are respotted.

Ice plant

When the outside temperatures were mild, reefers were simply ventilated, with the roof hatches propped open to catch the breeze as the train rolled along. More often, produce shippers took advantage of the many optional cooling services that were offered to keep the fruit in good condition during its trip east. Thus, an ice plant is another important industry on my layout.

13-8

Perishable service requires constant switching to keep a supply of iced empties moving back to the packing houses while loaded cars head east.

Refrigerator car fleets in 1950

Company	Cars
Pacific Fruit Express	38,840
Santa Fe Refrigerator Despatch	14,514
Fruit Growers Express	12,063
American Refrigerator Transit	11,457
Merchant's Despatch (NYC)	9,690
General American Transportation	6,122
Union Refrigerator Transit	5,726
Western Fruit Express	5,265
Armour Car Lines	4,198
Northern Refrigerator Co.	4,064
North Western Refrigerator Car Co.	3,398
North American Car Corp.	2,550
Burlington Refrigerator Express	1,895
Railway Express Agency	1,279

The PFE had its own ice plants in busy locations and contracted with independent ice companies in smaller towns. An icing deck kit by Creative Model Associates follows the PFE standard design. I kitbashed a pair of the firm's large kits into a single long deck, so I could ice an entire string of cars.

Located next to a big yard, the deck takes up one side of the layout, but it looks right since these facilities were often quite long. The one in Los Angeles was 30 cars long and double sided. Mine is eight car lengths long and single sided, but that's plenty big for my model railroad.

I used a Walthers cold storage building for my ice plant. Most of the actual plants were cast concrete with raised concrete beams and pilasters. I scratchbuilt the big cooling towers on the roof out of sheet styrene, since that's the spotting feature that indicates it's an ice-making plant and not just a storage building.

Icing the cars

I added a connecting platform to bring the blocks of ice to the conveyor chain on the deck. The ice rode on top of the chain, was broken up, and then placed into a car's ice bunker. Ice could also be stockpiled on the deck by covering it with burlap, which the deckmen did in anticipation of a train's arrival. On the deck, you might also occasionally see sacks of salt. The salt was used to make the ice even colder.

My ice deck sits alongside the yard on its own double-ended siding, parallel to the main line. I can run an entire train into the deck track to ice or re-ice the cars and then pull back onto the main without backing up or switching.

Most ice decks functioned in a similar manner to ice cars in transit, even in non-PFE territory to the east. Ice decks were located on the New York Central and the Pennsylvania to re-ice loads as necessary, often every day of a typical 14-day journey in the summer months.

To simulate activity at the ice deck, I always keep a few icing platform aprons in the down position with choppers, passers, and carmen hard at work, **13-7**. Blocks of ice travel from the ice plant down a conveyor chain to where they're needed, and the deckmen pull them off. Choppers begin breaking up the 300-pound blocks into big chunks, which the passers slide over sturdy boards to carmen working on the ice hatches. Then the carmen chip and pound the ice into the bunkers until about 10,000 pounds fill each one.

To make the deckmen, I modified Preiser construction figures and made my own ice tools from brass wire. Several manufacturers make clear blocks of ice, and I made the planks from three pieces of stripwood. Be sure to leave some planks and broken blocks of ice on the ground to represent what falls from the car tops during the icing process.

13-9

There wasn't much time to clean the exteriors of PFE's reefers during the busy packing season, so they ranged from freshly painted to just plain filthy.

During my operating sessions, the yard switcher pulls empty reefers off a storage track and spots them at the ice deck for icing. Then these cars go into a local that delivers them to the packing plant for loading. A short time later, the same local gathers up loaded cars and returns them to the yard, so they can be switched into eastbound trains, **13-8**.

These cars may travel east as a group or block of cars in priority trains, or if there are enough cars, an entire train of perishables may be dispatched on a priority schedule. Either way, the trains make a trip around the railroad and disappear into a hidden staging area which represents all points east. When the cars reappear, I consider them to be empties ready to repeat the cycle.

Ice refrigerator cars

Most railroads handled at least some perishables that were either passing through or terminating locally, so PFE refrigerator cars were commonly seen nationwide. Modelers have always had HO PFE reefers, but some of the early ones need work to bring them up to snuff. Even so, today's modelers will find quite a few car kits and ready-to-run models that look great.

In the 1950s, steel refrigerator car designs were being refined, while PFE continued its older-car upgrading programs that began in the late 1930s. Refrigerator cars had been built of wood because of its insulating qualities, but better types of insulation made it possible to build new cars of more-durable steel.

By the 1950s, the changeover was in full swing, so you can model both wood and steel reefers. That alone makes reefer blocks more interesting, but you can take it a step further by modeling rebuilt cars as well. PFE made many subtle changes and improvements, such as the addition of air circulating fans or new styles of hatches.

Putting on the miles

During my study of old photos, I noticed the wide extremes of weathering on the prototype cars. Some were freshly painted, others were pitch black with soot and grime, and many fell between these extremes, **13-9**. (To capture this appearance, see my article "Weathering a little at a time" in the July 2005 issue of MR.)

Remember that when packing season was in full swing, there was little time to clean car exteriors. Switchmen just wiped away grime, so they could read the reporting marks.

To simulate this look, I've developed all sorts of weathering techniques, including wiping away weathering with a cotton swab. Don't get too carried away with weathering because the filthy cars were always mixed in with freshly painted cars and everything in between. And when they're all put together in a train, you've captured the look of perishable shipping on the PFE in the 1950s.

Quick and easy citrus groves

I used 2" foam spheres to form orange trees. After slicing off the bottom of each sphere, I inserted a dowel into the base as a tree trunk. The dowel anchors the tree and should only be a few inches long.

Select a pleasing shade of green to paint your trees, but be sure you choose a water-based paint that won't react with the foam.

Once the paint dries, spritz the foam sphere with spray adhesive. While the tree's surface is tacky, hold it over a box and shake on Woodland Scenics medium green coarse turf.

One focal point of my HO layout is the little town of Arroyo on the outskirts of Los Angeles. As I modeled it to represent the early 1950s, the crops of choice in the surrounding area are citrus fruits, oranges mostly, but lemons as well.

Citrus trees, especially when viewed from afar, appear to be nearly round. Rows of them look more like green balls than trees.

An easy tree technique

In modeling my groves, I used two different sizes of foam spheres to give my acreage a little variety. Spheres with a 2½" to 3" diameter look appropriate for mature HO scale trees, while 2" or smaller spheres are well suited for younger groves.

While mature citrus trees, especially oranges, look quite round from a distance, they don't look like lollipops on a stick, **1**. Use a knife or saw to cut varying amounts off the bottoms of the spheres, so your trees won't all sit at exactly the same height.

For trunks, I cut short lengths (not more than a few inches long) of dowel, added dabs of glue to the ends, and then pushed them into the bases of the spheres. I didn't paint the trunks since they wouldn't be seen anyway.

One defining characteristic of citrus trees is that they keep their foliage to the ground. In fact, citrus tree trunks will sunburn if they aren't shaded by leaves or special wraps. For modeling purposes, I simply used the dowels to anchor the trees and hold them a few scale inches off the ground.

I spray-painted the foam using flat, green paint, **2**. Be sure that the paint is foam-compatible (water-based) or else it may pit or dissolve the sphere on contact.

After the paint had dried, I covered a large work surface with newspaper and then spritzed the spheres with a spray adhesive.

While the adhesive-coated sphere was tacky, I held it by the trunk and sprinkled on Woodland Scenics medium green coarse turf, **3**. A box catches excess turf for re-use.

To thoroughly secure the turf to the spheres, I also sprayed them with hair spray. Try not to get the turf too wet or it will mat and lose what little depth it has.

Over another box (again for recycling purposes), I sprinkled on the fruit and then quickly spritzed it with hair spray, **4**. Don't linger with the spray, as the oranges' color will run if they get too wet.

Planting the crop

I first determined where the trees should be planted. Then, using a rule as a guide for even rows, I laid out the trunk locations, drilled the holes, and sprinkled a little Woodland Scenics yellow grass around each to represent fallen leaves. I also added some oranges as fallen fruit. I put the trees in the holes but didn't glue them in place, so I can remove them for cleaning.

I added a few finishing touches to liven up the scene, **5**. There are stacks of wooden picking boxes, three-legged wooden orchard ladders (each made from a piece of styrene and Central Valley ladder sprue), and concrete irrigation pipes (made from styrene tubing).

Some of my favorite details are the smudge pots, tools used to help keep a hard frost at bay. You can make your own, or they are available from Showcase Miniatures (P.O. Box 753, Cherry Valley, CA 92223).

I had left room for the groves in several key spots around Arroyo. The groves frame the town nicely and give it an agricultural look. They also help soften and disguise the edges of the layout.

4

Use hair spray to secure the turf and then sprinkle on the oranges. Once the tree is loaded with fruit, spritz it again.

5

Once the trees are in place, you can add ladders, picking boxes, workers, smudge pots, and vehicles to add interest to the scene.

FREIGHT TRAINS

14-1

CHAPTER FOURTEEN

The 1950s railroad scene

By Robert S. McGonigal

The richness of the 1950s railroad landscape is apparent in this view at Crawfordsville, Ind., of a New York Central steam train among a water tower, code-line poles, trackside shanties, diamond crossings, and a single-story interlocking tower.
Linn H. Westcott photo

Compared with the stripped-down physical plant of today, the railroad environment of the 1950s was extraordinarily rich. A multiplicity of structures and equipment, from grand passenger stations to tiny shanties, and hulking locomotive coaling plants to slender signal masts, populated the railroad landscape, **14-1**. Most of the hundred-plus railroads in operation in the '50s designed and built their own structures, resulting in great variety wherever two or more carriers came together.

The most basic of the industry's elements – track – is indicative of the richness of the 1950s railroad scene. For one thing, there was a lot more of it. Although rail route-mileage peaked in the United States around 1915, there were still about 225,000 miles in operation in the early '50s; today, the figure is 100,000. Track miles – the total of all main tracks, sidings, spurs, and yards – stood at 379,000 in 1955 versus 169,000 today, an indication of the greater prevalence of double- and multiple-track main lines in the '50s, as well as the higher number of yards and industry tracks.

Most lines consisted of only a single track. When trains had to meet each other, one would go into a siding. The busiest main lines, mostly east of the Mississippi River, had more than one main track. Double track (one for each direction) was the most common multi-track arrangement, but traffic density could require three, four, or even six tracks. Multiple-track mileage decreased though the 1950s, partly because of the general decline of the industry but also because of the expansion of centralized traffic control (CTC) systems.

Introduced around 1930, CTC placed turnouts and signals of a given territory under the direct control of the dispatcher, who under earlier systems had to relay instructions to operators along the line, who would then prepare and deliver train orders to operating crews.

The development of CTC greatly increased track capacity, allowing single-track roads with growing traffic, like Western Pacific, to handle it without adding a second track. By the same token, roads with stagnant or declining traffic could remove tracks; thanks to CTC, hundreds of miles of New York Central main line were slimmed down from four tracks to two. During the 1950s, route-miles with CTC more than doubled, from 13,300 in 1950 to 29,100 in '59.

Welded rail was just coming into use in the 1950s, meaning that virtually all track was laid with jointed rail, the stuff that gave trains their distinctive *clickety-clack*. Rail was lighter, too. In 1955, nearly 40 percent of mainline track consisted of rail weighing less than 100 pounds per yard. Exposed tie ends, even on main lines, indicated that ballast was used more sparingly than it is today, though in the early '50s at least, track crews on some roads still manicured the ballast edge into a perfect line, with hardly a rock out of place, **14-2**.

On the other hand, financial belt-tightening in the late '50, especially on some eastern roads, led to the appearance of the weeds, low joints, and muddy ballast that would characterize some lines in the 1960s and '70s.

Extending from main and branch lines was a profusion of sidings and spurs. Many served lineside industries that depended on rail service for the delivery of fuel and raw materials and for the shipment of their products.

A B&O Mikado hustles a freight along jointed double-track bordered by knife-edged ballast and protected by semaphore signals. Linn H. Westcott photo

Such trackage ranged from a single track long enough for one or two cars to small yards.

To handle all this local business, railroads maintained numerous satellite yards in addition to major classification yards. Many small-town depots had a siding or two, including a house track that passed behind the station building.

Track was not confined to dedicated railroad rights-of-way. At many hundreds of locations across the country, circumstances forced trains to share city streets with cars and trucks, streetcar-style. Though not extinct today, street-running was much more common 50 years ago.

Structures for people

Structures were perhaps the most fascinating aspect of the 1950s railroad scene. They could be found anywhere there was a railroad, in an endless variety.

Passenger stations, where most of the official public contact with railroads took place, were among the most visible structures, **14-3**. They could be as simple as a three-walled shelter or as elaborate as a big-city union terminal.

Most numerous were small-town stations, which housed a waiting room, agent/operator's office, baggage room, and ancillary spaces in a few thousand square feet of area. Some included living quarters for the agent.

As a general rule, station size varied with the community's population – the bigger the town, the bigger the station. Exceptions would occur if a town was at the crossing of two or more lines or if it happened to host the headquarters of one of the railroad's operating divisions; in these cases, as at Durand, Mich., small towns would have substantial stations that housed offices in addition to the passenger facilities. Some served the trains of more than one railroad, making them junior-sized union stations without the imposing exterior and vast spaces of their bigger counterparts.

14-3

Variety in passenger-station design was nearly infinite. By the 1950s, as service waned, many stations, as this one at Round Lake, N.Y., on the Delaware & Hudson, began to show signs of age and neglect. Ladislav Dejnozka photo

14-4

Many grade crossings were once guarded by manually operated gates controlled by a watchman in an elevated shanty. Delaware & Hudson photo

The very earliest passenger stations were by and large replaced by new structures later in the 19th century, while the end of railroad expansion and the decline of passenger traffic resulted in the construction of relatively few depots after the 1920s. This meant that in the 1950s most stations were 30 to 80 years old. It was not uncommon for the original building to be altered, or sometimes moved, for use as a freight house.

Many smaller stations on a given railroad were of a common design developed by the carrier's engineering department, but some were unique, architect-designed structures. Construction ranged from wood frame to brick, stone, or concrete. Materials and architectural style often reflected local resources and preferences.

Structures for locomotives

Among the most distinctive railroad structures were those devoted to the servicing and maintenance of steam locomotives. Diesels required a different assortment of structures, making an engine terminal that hosted both types of motive power, as many did in the 1950s, a rich mix of old and new.

A large steam-engine terminal would have a roundhouse and turntable, a power plant, ash pits and disposal apparatus, a coaling tower, a water tank, a sand tower and storage area, an oil house, a machine shop, a foreman's office, and other buildings.

When diesels arrived, they were sometimes accommodated in purpose-built rectangular buildings with tracks running through them or simply serviced in the roundhouse. They also needed new fueling facilities, fed by oil storage tanks.

Major maintenance shops changed relatively little during the steam-to-diesel transition, except that they tended to get smaller, as less space was needed for even the heaviest diesel work. The diesel's reduced maintenance needs also resulted in the closure of many steam-era back shops.

Perhaps the structure that was most common in the 1950s, but the rarest today, is the water tower. Steam locomotives used vast quantities of water, so railroads placed water towers, built first of wood and later of steel, at frequent intervals along their lines. Even backwater towns along the railroad in the middle of nowhere had water towers.

To avoid having to stop trains so their locomotives could take on water, a handful of railroads, notably the New York Central and Pennsylvania, installed track pans on their busiest main lines. Pumping stations beside the tracks would fill the pans with water, which scoop-equipped locomotive tenders would draw upon as they passed at speed.

Somewhat less common were coaling towers. Built of timber, steel, or concrete, coal docks could be monstrous affairs spanning several mainline tracks.

Other structures

Another distinctive structure of the era was the interlocking tower. These small buildings, generally two stories tall, contained the control apparatus for signals and turnouts at junctions or other places where tracks came together. On the upper floor were the control levers, which were linked to the interlocking machine on the ground floor, which in turn was connected to the track and signals. These ingenious mechanical devices allowed the operator to line up routes while preventing him from setting up a potentially collision-causing combination – the machine interlocked all the equipment to prevent an accident.

Many interlockers were of the mechanical (or "armstrong") type, with all of the elements connected by steel rods, which sometimes required lots of muscle to operate. Later designs employed electricity and/or compressed air.

During the 1950s, numerous interlockings were "remoted" by having their functions added to another tower. Centralized traffic control enabled railroads to do away with towers altogether. Even so, many hundreds of them remained as the 1960s dawned.

Other less-prominent building types abounded. There were section shanties every few miles in which track workers stored tools, supplies, and motorized handcars. Many grade crossings were protected by gates controlled by watchmen in tiny shanties that were often built on stilts for a better view, **14-4**. Watchmen who stood in the street with a stop sign when a train was approaching worked out of ground-level shanties.

While radio communication was gaining acceptance in the 1950s, railroads still provided lineside telephones for employees to contact the dispatcher or other officials. Sometimes housed in a small box on a pole, phones could also be found in small phone booth-like buildings.

There were special facilities on routes with heavy perishables or livestock traffic. Most 1950s refrigerator cars were ice-cooled, so icing plants were provided to produce ice and for loading it into the reefers' bunkers. Many small livestock pens existed to hold animals for loading into stockcars; larger pens allowed the unloading of entire trainloads of stock.

14-5

A Louisville & Nashville tower at Winchester, Ky., is accompanied by a searchlight signal, gooseneck lamp, and wood stairs over tracks. Philip R. Hastings photo

Signals

Lineside signals of the 1950s came in a variety of styles: lower-quadrant semaphore, upper-quadrant semaphore, color light (grouped vertically, horizontally, or in a cluster), searchlight, **14-5**, position-light, and color position-light. Though most roads viewed the color light and searchlight as state-of-the-art, upper-quadrant semaphores were the most common type. Lower-quadrants could be found across the vast Southern Pacific and Union Pacific systems. The Pennsylvania and Norfolk & Western (which the PRR controlled) were the major users of position-lights, which employed rows of three amber lights to display signal aspects. Baltimore & Ohio liked the color position-light type.

Signals were mounted on masts on the right side of the track (generally in pairs, one on each side) or directly over the tracks they governed by means of structures not unlike those used today for signs on multilane highways. Chicago & North Western, which ran left-handed, needed signal bridges even in two-track territory, so its engineers, who sat on the right side of the cab, could see them.

At grade crossings, signals to warn road traffic of an approaching train were far more varied than is the case today. Twin flashing red lights beneath a "RAILROAD CROSSING" crossbuck, often accompanied by automatic gates, was a widespread design, but many variations existed within this basic type. Most common of all was the simple crossbuck, although details such as the angle of the X and the material and finish of the support post could vary significantly from railroad to railroad.

Signals of another type warned employees on top of boxcars – which in those days were still equipped with rooftop running boards – of overhead obstructions. These "telltales" consisted of a slender mast and a bracket from which was suspended a row of cords over and perpendicular to the track. If a brakeman facing away from the direction of movement felt these cords, he knew to immediately drop to a prone position; if he remained standing, the next – and probably last – thing he felt would be a bridge or tunnel portal.

The most common of all railroad structures were the millions of wood poles that carried communication, signal, and power lines alongside the right-of-way. Pole lines are rare today, but in the 1950s, they were as sure a sign of the presence of a railroad in the distance as the smoke from a steam locomotive.

Some might refer to the railroad scene today as sterile and bemoan the fact that modeling railroads involves little more than laying track. By contrast, the railroads of the 1950s featured structures of all shapes and sizes everywhere at trackside. It should really be no surprise, then, that the 1950s are the single most-popular period with model railroaders.

CHAPTER FIFTEEN

Modeling the right-of-way

By Bill Darnaby

15-1

Railroads in the 1950s offered a rich array of lineside details to model. Shown here are some of the more common classic-era features that you could add to your model railroad's right-of-way.
Photos by the author

The 1950s railroad right-of-way is an appealing modeling subject for a good reason – there are many interesting details to model, **15-1**. If you look at photos from the era, you'll see details alongside the tracks that are missing from today's railroad scene. As steam locomotives were being replaced with diesel-electrics, railroads, for a brief time, needed an infrastructure to support both types of power. Although new operating practices, such as radio communications and modern signaling, were changing how a railroad moved its trains, many manned towers and stations remained.

While great for the enthusiast and modeler, the 1950s weren't so great for the railroads themselves. Caught in the vise of rising costs, increased competition from trucking, and regulated rates, America's railroads were heading toward a two-decade period of decline that would leave few railroads unchanged. The slump finally ended with deregulation in 1980.

Although this period of financial hardship necessitated the changes that ultimately lead to the rebirth of the railroad industry, it caused many of the features modelers hold dear to be replaced by the spartan, efficient railroad scene of today. In contrast, the 1950s railroad scene was rich in visual appeal.

Many of the following right-of-way details are simple projects that can be completed in one or two evenings. After doing several, it won't take long to have authentic-looking scenes on your layout.

Track and ballast

When developing the right-of-way for my HO scale Cleveland, Indianapolis, Cincinnati & St. Louis RR (the Maumee Route), I was looking at period photos, and one thing that struck me was how neat railroad rights-of-way used to be. The roadbed structure consists of a subroadbed of compacted earth with well-defined drainage ditches on either side. This subroadbed was typically covered with cinders to keep down vegetation and promote good drainage.

The ballast roadbed was gravel or stone laid to a measured depth. Sometimes there were two layers, with a sub-ballast of cinder or gravel and a main ballast of stone, **15-2**. Railroad engineering textbooks provide very precise dimensions for right-of-way grading and ballasting.

Welded rail technology was just getting started in the 1950s, so most rail was still jointed with bolted connections. Because the rail had to be carried in cars of the period, the standard rail length was 13 yards (so they would fit in a 40-foot car). The rails' bolted joints showed prominently along the track. At this time, six-hole joint bars were common, but many main lines still had four-hole joint bars.

The rails were laid on creosoted wood ties. Overall, the ties were in better condition and more evenly spaced than they are today, particularly on secondary tracks.

Maintaining the track took a lot of manual labor by the section gang. A section was typically 10 miles of track, and the foreman and his section gang were required to patrol it daily, looking for and correcting problems.

In addition to section gangs, there were also bridge gangs that maintained a variety of bridges. Bridges were routinely and frequently inspected and painted to avoid deterioration through corrosion. A common feature of bridges of this period was the proud display in paint of the railroad name and, sometimes, the slogan or advertisement of passenger or freight service.

15-2

On rights-of-way, the edges of the ballast were neat and straight, providing a clear separation line against the underlying cinders.

15-3

Each section featured a storage shed, such as this speeder shed, and a track car turnout, where cars could easily be pulled off the track.

Sheds and structures

Railroads built a variety of structures to store the gangs' tools and equipment. Some were built specifically for that purpose, while others, such as old car bodies, were modified. In the days before hi-rail vehicles, workers from all departments – roadway, communication, signal, and mechanical – traveled the line on motor cars (speeders). Each department had its own fleet of cars, and each section had its own sheds for motor cars and tools, **15-3**. Other departments kept sheds and buildings at terminals or division points depending on how the work territories were divided.

Speeder sheds and other small buildings were usually built to a standard design and painted in a uniform color scheme, although there was a lot of improvisation. You can find plans for such structures in *Model Railroader* or from railroad historical societies.

Other miscellaneous structures along a right-of-way included fuel (both coal and oil) storage sheds near depots and offices, phone booths, outhouses, and pump houses by water tanks.

An active depot could feature many elements, including a train-order signal (in front of the depot), flatcar loading ramp, freight house, water tank, and pump house.

Along single-track lines, line poles commonly held two or three crossarms, while those along main lines could have as many as six or seven crossarms.

Depots

The one right-of-way structure everyone thinks of is the depot, and in the '50s, most depots still saw active use, **15-4**. In addition to being a place to pick up and drop off passengers, a depot was the railroad's office for that town. Local businesses and shippers worked with the station agent to ship and receive goods in their area.

Many railroads also used their depots as train-order offices. A train-order signal indicated whether or not the operator had orders from the dispatcher for directing train movements. Train-order stands (for holding pickup orders by trains on the move) and baggage carts were fixtures in front of a depot.

Line poles and towers

The advent of radio communication had a large effect on the appearance of the railroad right-of-way. Once radio became accepted for dispatching in the later 1960s, railroads started closing their depots. One of the biggest visual changes brought about by radio, however, was the loss of communication lines strung up on poles. In the 1950s, line poles followed practically every right-of-way and carried signal lines, telephone and telegraph lines, and low-voltage power lines, **15-5**.

Like many other railroad functions of the 1950s, signaling, particularly at interlockings, still operated manually. As a result, signal towers remained in daily use. Crossings or junctions would have a tower or small building to shelter the equipment and the operator. Towers were built from wood, brick, or concrete. Like depots and sheds, towers usually followed a railroad's standard design and decoration.

Towers built before 1920 were almost always constructed as mechanical interlockings in which the logic and actuating mechanism was made up of a series of mechanical levers, cranks, and rods. The bulk of this mechanism was outside the tower and very visible along the right-of-way.

After 1920, electric and pneumatic actuating mechanisms for track switches and signals became popular. Some old mechanical plants were converted to this style, but the bulk of mechanical plants remained intact through the 1950s (a few even made it into the 1990s). The increased use of centralized traffic control and automatic signaling saw the steady replacement of manned towers at the end of the decade.

Signals

Signals used in the '50s look different than they do today. Because there were so many railroads (nearly 100 Class 1 lines alone), there was a wide variety of signal types. Some were quite distinctive, like those of the Pennsylvania RR or Baltimore & Ohio, while others were variations on basic color-light or searchlight signals. During this period, you could often identify the ownership of a railroad right-of-way simply by the style and type of block signals found along its tracks.

Interlocking signals also tended to reflect the practices of the railroad that built and maintained the plant. It was common to find the standard signal of each railroad within an interlocking plant if a particular road preferred its own signal, **15-6**.

East of the Mississippi, the most common form of interlocking used in the early 1950s was the mechanical interlocking with lever-operated semaphore signals. During the decade, many such interlocking plants were modernized by replacing the semaphores with color-light signals. Switch points, however, might still be operated by levers. Interlocking plants that had

15-6

Both the Maumee's color-light signal (left) and the B&O's position-light signal (right) are in use where the two roads cross. The relay cabinet (center) is painted black as it would be on an eastern railroad.

15-7

This grade crossing is protected by flashing signals that are painted black and white and mounted on striped concrete piers.

15-8

Water and coaling towers were prominent right-of-way features. As railroads converted to diesel, they were removed.

been upgraded were easily identified by the empty signal-actuating rod guides along the track.

East of the Mississippi, signals and relay cabinets of this period tended to be painted black, mast and all. Out west, silver or aluminum signal masts and relay cabinets were already common, but that trend didn't catch on in the eastern states until the 1960s.

Grade crossings

Grade crossing protection in the 1950s was not as sophisticated as it is today because there was not nearly the amount of vehicular traffic. A simple crossbuck, usually made from old rail and wooden cross boards, was standard grade-crossing protection, and as with signals, some railroads had signature crossbucks, such as the cast-iron ones on the Pennsy. A common practice of the period was to paint the railroad name on the crossbuck and sometimes stripe the pole in contrasting colors, usually black and white.

Flashing crossing signals were used only on roads with heavy traffic, where vehicles ran at high speeds, or where visibility was poor, **15-7**. Crossing gates were reserved for the most heavily trafficked roads, and automatic gates were just becoming common. Where there was a lot of train traffic, particularly switching, the gates would be manned, with an elevated tower for the operator.

Steam facilities

Facilities required to operate steam locomotives were very visible right-of-way details, **15-8**. They included water towers, water plugs, coaling towers, and in the West, oil columns.

Coaling stations could be found midway between terminals on many roads, depending on the coal capacity of the tender. Steam engines with larger-capacity tenders eliminated the need for intermediate stops, but smaller engines required frequent refueling.

As the 1950s progressed and railroads completed their conversion to diesel power, they removed coal and water towers. Concrete coal towers, especially those that straddled main-line tracks, were difficult to remove. As a result, some have lasted into the 21st century.

On roads that moved livestock, lineside engine water facilities frequently had low-level water plugs with nozzles for spraying livestock in hot weather.

CPSIA information can be obtained
at www.ICGtesting.com
Printed in the USA
FSHW04n1920270318
45968FS

9 780890 247181